Is Nuclear Right for My Family?

Making Energy Choices Around the Kitchen Table in a Climate-Changing World

Peter Ozemoyah

Published by Steady Current Press
An imprint of Zep Consult Inc.
steadycurrentpress.ca

First edition. June 2026

ISBN (paperback): 978-1-0678498-0-1
ISBN (eBook): 978-1-0678498-2-5

Printed in the USA

To All Who Seek Knowledge and Empowerment

Contents

Introduction

What's ahead

A Quiet Evening, Then Darkness

The first hint was the sudden silence.

"Dad, the Wi-Fi's gone!" my son called from the living room.

Before I could answer, the kitchen lights flickered and died. The hum of the refrigerator stopped. My daughter peeked around the doorway, wide-eyed.

"Is it just us?"

I checked the window. The whole street was dark. A winter storm had knocked out the power.

We gathered in the living room, wrapping ourselves in blankets while sleet tapped against the glass. My wife lit a candle, its small flame casting long shadows.

"How long will it last?" my son asked.

"Could be hours," I said, though I had no idea.

The questions kept coming. Where does our electricity come from? Who fixes it when it breaks? Why does one storm bring everything to a stop?

We talked as the candle sputtered. The kids drew our house on a scrap of paper and began tracing imaginary wires to distant power plants. What started as a cold inconvenience turned into a night of discovery. By the time the lights blinked back on after midnight, we knew more about our own lifeline than we ever had before.

Energy as a Family Story

That evening taught us something lasting. Energy is not just a technical system humming in the background. It is the quiet rhythm of family life. It shapes how we cook dinner, heat bedrooms, and connect with each other. Yet we rarely notice it until it falters.

Families everywhere share similar moments. A grandmother in Ontario remembers smoggy summers before coal was gone. A teenager in Kenya wires a solar lamp so homework can continue after sunset. In Germany, rooftop panels become a badge of local pride.

These stories reveal that energy is more than wires and kilowatts. It is memory and trust, anxiety and pride, the backdrop to every ordinary day.

What This Book Offers

This book invites you to see energy through a human lens, to step past statistics and into everyday life.

Across continents, we visit Canadian towns where nuclear plants quietly keep winter air clear and lights steady, German villages still debating the legacy of their nuclear exit, and South African neighbourhoods where rolling blackouts test both resilience and neighbourly care. In each place, energy is not just technology but part of the family story.

Across generations, children become teachers when they track household electricity use or build small solar ovens for school projects. Elders share memories of coal mines, hydro dams, and the first glow of electric light. Parents weigh the promise of renewables against the steady reliability of nuclear power, trying to balance climate urgency with security and cost.

You will meet welders who retrained to build reactors, students installing rooftop panels, artists transforming power lines into public sculpture, and activists standing in the rain for cleaner air. Their decisions — large and small — are shaping the energy systems of tomorrow and showing that the future is written not only in laboratories or parliaments, but around kitchen tables everywhere.

Fig. 1. Author's portrayal of boardroom and kitchen-table discussions of energy and energy bills

Why It Matters Now

The journey ahead is both daunting and hopeful. Climate change is real, fossil fuels still dominate, and injustices linger. Yet families today hold more tools and collective wisdom than any generation before. Looking ahead is not about certainty; it is about courage — and courage, like energy itself, grows stronger when shared.

The chapters that follow show that the energy transition is not decided only in parliaments or corporate boardrooms. It begins in kitchens and living rooms, in the habits we teach our children and the leaders we choose at the ballot box.

Every household is part of the global energy story. By understanding the systems that keep our lights on — and the trade-offs each option carries — we gain more than knowledge. We gain agency: the ability to choose a cleaner, fairer, more resilient future together.

So as you turn the page, imagine the soft glow of a candle on a stormy night, a family gathered close, and the quiet realization that the energy we share is as human as any story we tell. This is where our journey begins.

Part I

Families, Energy, and Everyday Life

Energy is not just about technology. It is also about how families keep warm, pay their bills, and live safe, healthy lives. This part brings energy down to earth by showing how radiation, electricity, and technology affect everyday life.

CHAPTER 1

What Keeps the Lights On?

Why energy matters in every family's life

The Invisible Thread of Everyday Life

Every morning before our eyes even flutter open, electricity is already at work. It hums through the wires behind our walls, trickles into the phone charging on the nightstand, powers the alarm clock that pulls us out of sleep, and brews the coffee that kick-starts the day. Energy is woven into the rhythm of family life.

Across the planet, this same quiet choreography plays out: In Lagos, a kettle whistles before sunrise; in Tokyo, bullet trains glide out of stations powered by intricate grids; in rural Peru, micro-hydro plants keep early-morning markets lit.

Most families don't think about it until something goes wrong: a blackout during a storm, a bill that's higher than expected, or news of an energy crisis half a world away that somehow raises local prices. When the lights flicker or the furnace stalls, urgency replaces habit. The 2021 winter storm in Texas, for example, left more than four million households without power and caused an estimated US$130 billion in damages, a reminder that reliability is never guaranteed. Similar "load-shedding" outages strike South Africa daily, forcing families to cook and study around scheduled blackouts.

This chapter pulls back the curtain on that invisible system. It answers a simple but powerful question: What really keeps the lights on? The answer involves more than just power plants and wires. It's a story of choices, trade-offs, and resilience — one that connects families in living rooms to global networks of mines, reactors, solar farms, and policy debates.

The Global Energy Mix: A Patchwork Quilt

Electricity doesn't come from just one place. Around the world, countries rely on different energy sources depending on geography, history, politics, and values. Think of the global energy landscape as a patchwork quilt: colourful, varied, and stitched together by local needs and global challenges.

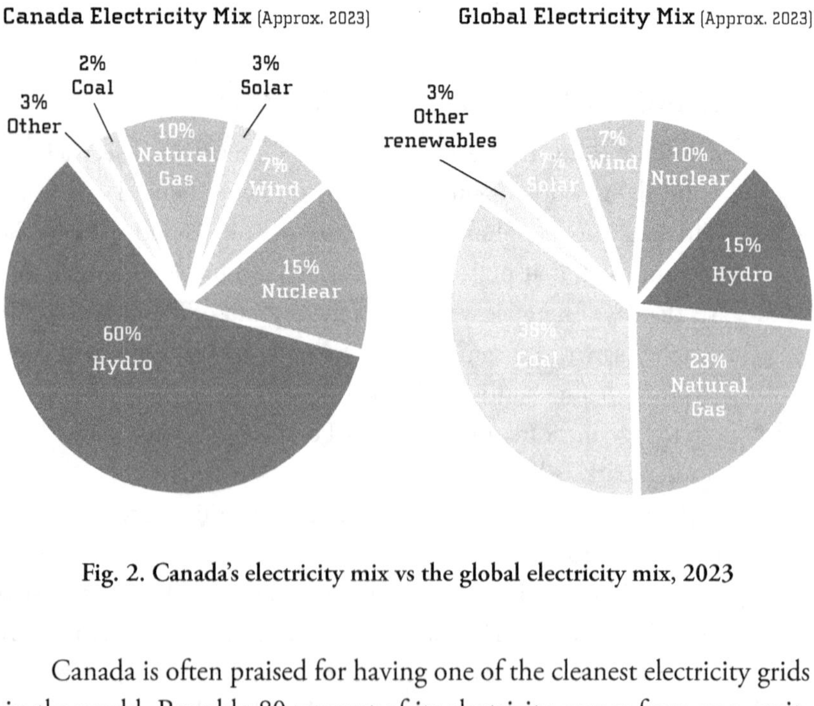

Fig. 2. Canada's electricity mix vs the global electricity mix, 2023

Canada is often praised for having one of the cleanest electricity grids in the world. Roughly 80 percent of its electricity comes from non-emitting sources like hydro and nuclear. In Quebec and British Columbia, rivers are harnessed by massive dams that together produce more than 200 terawatt-hours of power each year. In Manitoba, hydro dominates as well. In Alberta, however, fossil fuels still play a central role, even as wind and solar grow rapidly — wind capacity there doubled between 2015 and 2023. Ontario leans heavily on nuclear power — nearly 60 percent of its electricity comes from reactors that provide a steady, carbon-free baseload.

Germany chose a different path with its *Energiewende* — phasing out nuclear while ramping up wind and solar. Fields of panels and forests of turbines are now common sights. But there's a paradox: Without nuclear, Germany has leaned on coal and imported gas, particularly from Russia. When the war in Ukraine disrupted gas supplies, German families felt the squeeze through higher bills and greater energy insecurity. Household electricity prices briefly climbed above €0.40 per kilowatt-hour, among the highest in Europe.

France went the other way. By betting big on nuclear decades ago, it secured a stable, low-emission grid. French families now benefit from cleaner electricity and relatively stable prices that average roughly half those in Germany, even during fuel shocks.

The United States is vast and diverse, so no single model applies. California is a solar powerhouse, Texas leads in wind, West Virginia still leans on coal, and states like Illinois and New York maintain strong nuclear fleets. Families in one state may see rooftop solar panels as normal, while in another, coal plants still dominate the skyline.

Across rural Kenya, electricity is increasingly delivered not by distant power plants but through locally managed solar mini-grids. These cooperative-owned systems reflect the uneven, place-specific nature of the global energy mix, bringing reliable power to villages long overlooked by national grids and supporting everything from health clinics to small enterprises. Tuvalu, a small Pacific Island nation, illustrates the fragility of energy. Diesel generators provide most of its power, but imported fuel is expensive and vulnerable to shipping disruptions. Rising sea levels threaten both homes and energy infrastructure. For Tuvaluan families, energy security is inseparable from climate survival.

South Korea continues to invest in nuclear, both for domestic use and as an export technology. Strong safety standards and clear communication have built trust with families who see nuclear as a reliable foundation. Brazil has pioneered floating solar panels on hydro reservoirs, combining two clean sources without using new land. Families benefit from both stable electricity and environmental stewardship.

Each energy mix reflects local priorities: affordability, security, independence, and sustainability. There is no universal recipe — only choices that affect the air we breathe, the stability of our climate, and the size of our monthly bills.

Fossil Fuels: Familiar, but Flawed

Coal, oil, and natural gas powered the world into modernity. They fueled factories, heated homes, and made cars and airplanes possible. They remain deeply embedded in global infrastructure, providing dense, reliable energy.

But the drawbacks are increasingly clear. Burning fossil fuels emits carbon dioxide (CO_2), the main driver of climate change. Coal is the dirtiest, releasing not just CO_2 but mercury, sulphur, and particulates that cause smog, acid rain, and health problems like asthma. The World Health Organization estimates that air pollution linked to fossil fuels kills around seven million people each year. Natural gas is sometimes marketed as a "bridge fuel," but methane leaks make it a major climate threat. Methane traps over eighty times more heat than CO_2 over twenty years.

Then there's volatility. Families feel it directly. When oil prices rise due to global conflicts or supply disruptions, heating bills and gasoline costs spike. The energy shock that followed Russia's invasion of Ukraine offered a clear lesson in how fragile fossil-fuel systems can be. Within weeks, heating costs across Europe surged, while fuel prices in the United States spiked as well, briefly pushing gasoline above five dollars a gallon. What felt like a distant geopolitical crisis quickly became a household concern on both sides of the Atlantic.

Fossil fuels are familiar, but they are flawed — environmentally, economically, and politically. The challenge is not only phasing them out, but ensuring that families dependent on them for jobs and livelihoods are supported in the transition. Programs like Spain's coal-region "just transition" funds and Canada's federal retraining initiatives show that governments can cushion communities while cutting carbon.

Renewables: Clean and Growing, with Hurdles

Renewables — solar, wind, hydro, geothermal — are celebrated for being clean and sustainable. Once built, they can provide decades of power without emitting greenhouse gases.

Canada's geography favours hydro, with dams providing the majority of power in provinces like Quebec. Wind is growing in Alberta, Ontario, and Nova Scotia. Solar remains smaller but is expanding quickly as costs fall. Global solar module prices have dropped more than 80 percent since 2010.

Globally, renewables are now the fastest-growing part of the energy sector. In countries like Denmark, wind regularly supplies over half of the national demand. Solar farms in China and India have scaled up at breathtaking speed. China installed more solar capacity in 2023 alone than the entire world added in 2015.

But renewables face hurdles. They are intermittent — the sun doesn't always shine; the wind doesn't always blow. Without storage or backup, they can't meet constant demand. Large hydro dams can flood land and displace communities, and some offshore wind projects have raised concerns about marine ecosystems.

For families, renewables often feel visible and empowering. Rooftop solar gives households a sense of independence. Community wind projects can reduce bills and foster local pride. But renewables need partners, other energy sources that fill in the gaps when nature takes a break.

Nuclear Power: Steady and Misunderstood

Nuclear energy is powerful, clean, and reliable — but also misunderstood.

It works by splitting uranium atoms in a process called fission, releasing heat to produce steam that spins turbines. A tiny uranium pellet can power a household for months. Nuclear plants don't produce air pollution, and they run continuously, day and night.

Yet public concerns linger. Accidents like Chernobyl and Fukushima cast long shadows, even though today's reactors are far safer. Waste is radioactive, but it is small in volume, tightly regulated, and stored securely — unlike the invisible emissions from fossil fuels. Cost is high up front, but over decades, nuclear often proves competitive due to low operating costs.

Canada's CANDU reactors have a strong safety record and provide about 60 percent of Ontario's electricity. Around the world, new technologies like Small Modular Reactors (SMRs) are being developed. These compact, factory-built units are designed for added safety, flexibility, and use in remote communities. Finland's Onkalo repository shows how long-term waste can be safely isolated for millennia.

For families, nuclear's value is stability. When storms rage or demand spikes, nuclear plants quietly keep the lights on. During the 2021 Texas freeze, U.S. reactors in neighbouring states ran without interruption, helping stabilize regional grids.

The Grid: An Invisible Web

Generating electricity is only half the story. Getting it to your home is the other half. That's the job of the grid, a vast web of wires, substations, and transformers that balances supply and demand every second of every day.

When families crank up air conditioners during a heatwave, regional grid operators forecast demand and coordinate generation, reserves, and transmission in real time to help prevent blackouts. Much of the U.S. and Canadian electric system is highly interconnected — tied together through dozens of high-voltage cross-border lines — allowing electricity to flow between provinces and states where it's needed most, enhancing reliability and flexibility.

But much of the grid is old, built for a different era. As electric vehicles, smart devices, and renewable energy grow, grids need modernization. Families are already seeing changes through smart meters and demand-response programs that reward using energy at off-peak hours.

When the grid fails, the consequences are stark. The 2021 Texas blackout left millions of families freezing in the dark, revealing how fragile even advanced systems can be. In Puerto Rico, after Hurricane Maria, some communities went months without electricity. A resilient, fair grid is as essential as clean generation itself.

Families: From Consumers to Participants

For much of history, families were passive consumers of energy. But that's changing fast.

- Rooftop solar lets households generate their own electricity.
- Home batteries store energy for use during outages.
- Electric vehicles not only replace gasoline but can eventually feed power back into the grid.
- Smart thermostats help families save money while reducing strain on the grid.

Governments encourage participation through grants, tax credits, and co-op models. Canada's Greener Homes program supports families in insulating homes, installing heat pumps, or adding renewables. In the United States, the 2022 Inflation Reduction Act expanded similar credits, spurring a surge in residential solar installations.

Even small actions matter: teaching kids to turn off lights, sealing drafty windows, and upgrading old appliances. Together, these build a culture where families are active partners in shaping the energy system.

A Future with Many Options

So, what keeps the lights on? The answer is complex and evolving. Fossil fuels, renewables, nuclear power, and efficiency all play roles. The ideal future is not one single source, but a blend, tailored to local needs, balanced for fairness, and designed for resilience.

For families, this means not being locked into one fragile system. It means lower bills, cleaner air, more stable communities, and opportunities for children to work in industries of the future. Analysts project that by 2050, more than half of global electricity could come from renewables, with nuclear holding steady and fossil fuels declining sharply.

Energy is the lifeblood of daily existence, and the decisions we make about it will shape the kind of world our children inherit. Every family is already living inside the energy story. The question is: What role will we play in writing the next chapter?

Family Action:

- Find your local energy mix by checking your utility bill, your provincial or state energy website, or your community's annual energy report.
- Note the top two sources powering your home.

CHAPTER 2

What This Means for Your Family

The health and financial stakes of how we power our homes

Where Energy Becomes Personal

When families think about energy, it usually arrives as a bill — numbers on a page, an email notification, or an automatic withdrawal from the bank account. Hidden behind those digits are health, safety, comfort, and resilience. It determines whether children can sleep without coughing, whether food stays cold during a storm, and whether family budgets stretch beyond the basics.

Worldwide, household energy costs account for an average of 6 percent of monthly income, but in low-income communities, that share can rise to 20 percent or more, a reminder that energy decisions shape how families live, what they can afford, and how safe they feel.

This chapter takes the conversation out of power plants and parliaments and places it squarely in the kitchen, the bedrooms, the sidewalks of your neighbourhood. Because what energy means for the world matters, but what it means for your family is where the story begins.

Power, Prices, and the Budget That Holds Everything Together

Amanda and Kyle live in a small bungalow in Calgary with their two young kids. For years, they hardly thought about energy. The lights came on, the furnace kicked in, the bills arrived. But in March 2022, their monthly bill — usually around $120 — shot past $300. Nothing had changed in their home. The culprit was natural gas: Global prices had spiked because of supply shortages tied to the war in Ukraine. Suddenly, Amanda and Kyle were choosing between lowering the thermostat to a chilly 17°C (about 62°F) or dipping into their savings.

This is the reality of fossil-fuel dependence. Your bill is tied to events thousands of kilometres away. A refinery fire in Texas, a war in Europe, or an unusually cold winter can raise costs — and families feel it first. During 2022, natural-gas prices in North America briefly tripled. In Europe, they rose more than sixfold, creating ripple effects from grocery prices to mortgage stress.

Now look east to Quebec or Ontario in Canada, where electricity grids lean heavily on hydro and nuclear. Families there heat homes, run appliances, and charge cars with bills that remain relatively stable year to year. Why? Because hydro and nuclear plants don't depend on volatile international fuel markets. They produce power steadily, regardless of whether gas prices soar.

For families, stability is everything. It means predictable budgeting, less stress, and peace of mind. Yet stability is unevenly distributed. In the United States, nearly 30 percent of households struggle to pay energy bills, and more than 15 million face disconnections each year. In sub-Saharan Africa and South Asia, hundreds of millions of families still lack reliable electricity altogether. Energy poverty — spending a disproportionate share of income on energy, or going without — shapes the lives of billions. The International Energy Agency estimates that 770 million people remain completely without electricity today.

Energy costs are not just numbers. They are groceries forgone, medicine delayed, children studying in the dark. They are the difference between security and precarity.

Breathing Easier Isn't Just a Metaphor

In the early 2000s, Ontario's skies were often thick with smog. Emergency rooms filled with children suffering asthma attacks. Schools sometimes told kids to stay indoors. Pollution was invisible but relentless, creeping into lungs and shortening lives.

Then, Ontario phased out coal. Nuclear, hydro, and renewables filled the gap. By 2015, the province celebrated its first summer without a single smog advisory. Hospital visits for respiratory problems dropped. Children like nine-year-old Jacob in Mississauga used their inhalers less and played outside more. Similar improvements are seen worldwide when coal is replaced. After Britain's "coal-free" days began in 2017, the country recorded its lowest urban particulate levels in modern history.

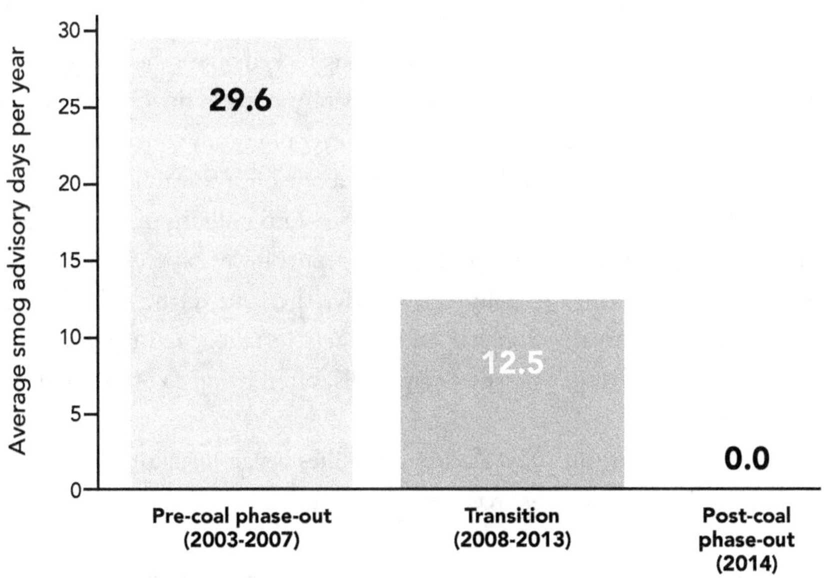

Fig. 3. Average number of smog advisory days in Ontario before, during, and after the coal phase-out

The lesson is clear: Energy choices show up in your family's lungs. Burning coal and gas releases particulates and nitrogen oxides that trigger asthma, heart disease, and premature death. The World Health Organization estimates that seven million people die prematurely every year from air pollution, much of it tied to fossil fuels.

When clean energy replaces not-so-clean fuels, families notice. They notice clearer skies, fewer doctor visits, children running outside without wheezing. Clean power is not just a climate win. It is an intimate, personal health victory.

Everyday Decisions with Bigger Impacts

Energy isn't only decided by politicians and corporations. Families shape it too, through everyday choices.

In Nairobi, families with prepaid electricity meters often run out of credit, plunging homes into darkness. New community-led solar micro-grids are providing steady, affordable power so children can study after sunset and parents can keep food safe in refrigerators.

In Halifax, Nova Scotia, the Mendes family rents an apartment where rooftop solar isn't possible. Rather than being locked out of the energy transition, they joined a community solar program outside the city, receiving monthly bill credits from panels they will never physically touch. For them, shared solar opened a door that ownership alone once controlled.

In Bangladesh, solar home systems have lit up millions of rural households, replacing kerosene lamps. Families report lower expenses, healthier homes, and more time for children to study. The same is true in Barbados, where almost every rooftop has solar panels installed, reducing dependence on the grid and lowering energy bills, often by up to 80 percent for some households.

In Texas, after the 2021 blackout, families began installing home batteries and backup systems, recognizing resilience as more than a luxury. It was survival.

Not every family can afford rooftop solar or a new EV, but even small steps matter: switching to LEDs, upgrading insulation, doing laundry or running dishwashers at night. These changes cut bills and emissions while building a sense of agency. If every Canadian household replaced just one incandescent bulb with an LED, the annual savings would power more than 250,000 homes.

Safety and Resilience When It Matters Most

The summer of 2021 brought a deadly heat dome to British Columbia. Temperatures soared past 40°C (104°F). Air conditioners strained, some grids faltered, and hundreds of people — mostly elderly or vulnerable — died of heat exposure in their own homes.

Extreme events are becoming routine. Hurricanes, floods, wildfires, and ice storms all threaten grids. Families discover quickly that energy isn't just comfort — it's survival.

In Texas, the 2021 blackout left millions freezing in the dark. Families burned furniture for heat, hospitals struggled to power ventilators, and dozens died of hypothermia. In Puerto Rico, after Hurricane Maria, many families lived without power for months. Hospitals ran on fragile diesel generators. Children studied by candlelight. In India, heatwaves push grids to breaking points, shutting down fans and refrigerators when families need them most.

Resilience means systems that bend but don't break. It means distributed solar with storage, microgrids, diverse supply sources, and strong baseload power from nuclear and hydro. For families, resilience is peace of mind: knowing the fridge will stay cold, the heater will keep running, the lights will come on even in crisis. Studies show that every dollar invested in grid hardening or home backup systems saves several dollars in disaster recovery.

Raising Energy-Aware Kids

Children today grow up hearing words like "climate crisis," "wildfire smoke," and "blackouts." They see solar panels on rooftops, wind turbines on road trips, and images of flooded cities online. They are not shielded from energy's realities.

In a grade five classroom in Kingston, Ontario, Canada, students studied their household bills for the first time. They tracked appliance use, learned where electricity in Ontario comes from, and debated clean energy options. Samira, age ten, was so inspired that she asked for a solar science kit for her birthday.

Around the world, young people are leading. In New Delhi, students host debates on nuclear and renewables. In Kenya, girls designed solar charging stations to power their schools. In California, teens built apps to track energy use at home. International surveys show that more than 70 percent of teenagers view climate change as "very important" to their future.

Parents can nurture this by:

- Naming what powers appliances.
- Talking openly about bills and choices.
- Trying hands-on projects like solar ovens or wind-powered toys.

Kids who understand energy see it not as mysterious, but as something they can shape. They grow into citizens ready to make informed choices.

The Hidden Costs Families Carry

Not all energy costs show up on the bill. Families bear hidden burdens:

- Health costs: doctor visits, medications, hospital stays for asthma or heart disease caused by pollution.
- Time costs: hours lost to power outages, waiting for food or water deliveries.
- Emotional costs: anxiety about bills, fear during storms, frustration when systems fail.

When policymakers debate "cheap" energy, families often carry these invisible expenses. A coal plant may produce cheap electricity per kilowatt-hour, but families nearby pay with their lungs. A gas grid may seem flexible, but when prices spike, families pay with their wallets. True affordability must account for these hidden costs. Economists estimate that if the health damages from coal pollution were priced into electricity, the real cost per kilowatt-hour would more than double.

From Personal to Collective

Energy begins at home — in the bills, the lungs, the routines of families. But it doesn't stay there. Collectively, family choices shape markets, influence policies, and create cultural norms.

When millions switch to efficient appliances, utility companies adjust. When parents support clean energy policies, governments respond. When children learn to care about energy, they grow into innovators, voters, and leaders.

Energy is not distant. It's as close as the plug in your wall, the breath in your lungs, the warmth in your child's bedroom. Families already live inside the energy story. The question is whether we'll remain passive consumers or active participants shaping a future that is cleaner, fairer, and more resilient.

Family Action:

- Track your household energy use for a week and discuss it with your children.
- Explore local programs for rebates on insulation, heat pumps, or solar.
- Join or support community energy projects if available.
- Talk to neighbours about resilience: What happens if the grid fails tomorrow?
- Teach kids that energy is not just science, but also fairness, health, and imagination.

CHAPTER 3

Radiation in Perspective

Everyday exposures, myths, and realities

Separating Perception from Reality

Radiation has a way of capturing the imagination. For many families, the word itself stirs unease. We picture glowing barrels stacked behind fences, warning signs plastered in yellow and black, or news clips of accidents where invisible particles seemed to creep through the air. Popular culture has amplified these fears with stories of superheroes mutated by radiation or wastelands poisoned beyond repair. But the reality, when examined closely, looks very different.

In truth, radiation is measured, monitored, and well understood. Doses are quantified in millisieverts (mSv), and health agencies across the world set exposure limits for workers and for the public with large safety margins.

Radiation is not an intruder in our lives; it is a constant companion. It flows from the sky, the earth, the food on our tables, and even from our own bodies. We live with it every day, often without realizing it. Understanding this is the first step to seeing radiation not as a monster under the bed, but as a natural phenomenon that deserves perspective.

On average, people receive about 2–3 mSv each year from natural background sources alone, and in some high-altitude or high-granite regions, that background can be 5–10 mSv without any observed health effects in residents.

Radiation in Everyday Life

Families already encounter radiation in ordinary routines. The fruit bowl on the counter is a perfect place to start. Bananas, rich in potassium, contain naturally occurring radioactive isotopes. So do nuts and beans. Every time your child eats a banana, they absorb a tiny dose of radiation. It is harmless — but measurable.

A single banana corresponds to roughly 0.1 microsievert (μSv), a fraction of a millionth of a sievert. You'd need to eat tens of thousands of bananas in a short time to approach even a routine medical X-ray.

Step into a kitchen and look at the countertop. If it is made of granite, it too releases small amounts of natural radiation. Nothing alarming, but it is a reminder that the materials we live with carry traces of energy from the earth itself.

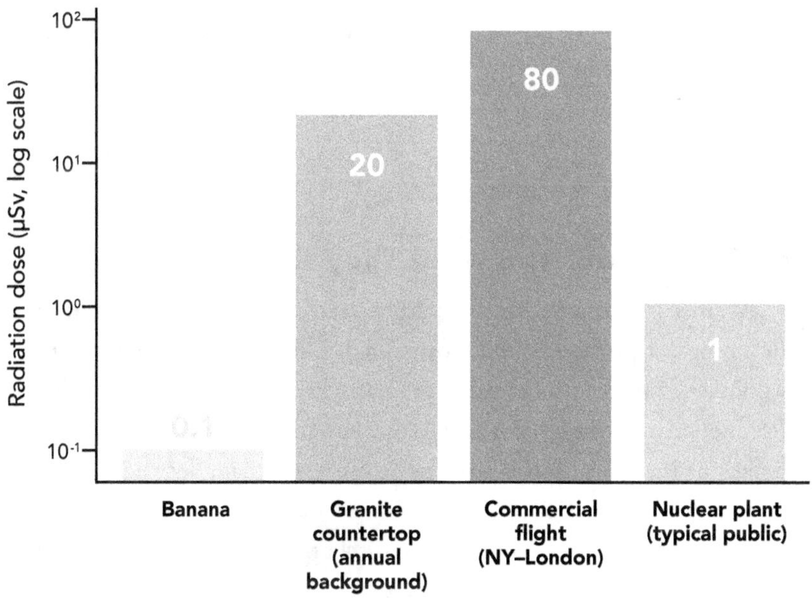

Fig. 4. Comparison of typical radiation doses from common sources

Granite counters typically add far less than 1 mSv per year to exposure, usually so low that sensitive instruments are required to detect it.

Then there is travel. A family flying across the Atlantic receives more radiation from cosmic rays at high altitude than they would in a year of living next to a nuclear power plant. Pilots and flight attendants, who spend years working at altitude, actually accumulate higher doses than most people who work in the energy industry.

A single transatlantic round trip is commonly in the range of 0.05–0.1 mSv, and frequent fliers can accumulate 2–5 mSv per year from flights alone. In some countries, aviation workers are categorized as "radiation workers" for this reason.

And, of course, the most familiar exposure of all comes from the sun. Families love the warmth of a summer day, but ultraviolet rays are radiation too. We don't run from the sun. We take precautions with sunscreen and hats. We manage the risk rather than fearing it absolutely.

This everyday approach — recognize the source, understand the dose, take simple precautions — translates well to all forms of radiation.

Radiation is part of the background of life on Earth. We cannot escape it, but we can understand it — and once we understand it, it stops being mysterious.

Perspective grows when we compare: Background radiation for most people over ten years (20–30 mSv) is still below a single high-end medical imaging course like multiple CT scans.

Nuclear Plants and What They Really Release

Against this backdrop of natural radiation, the contribution of nuclear power plants seems almost trivial. Many people imagine reactors as unsafe, with radiation, which are particles that can't be seen slipping into the surroundings. In reality, the radiation released under normal operations is extremely small — far below levels that affect health.

Regulators set strict public limits near plants — typically 1 mSv per year above background — and actual discharges are usually a tiny fraction of that limit.

A family living near a nuclear plant will receive, on average, less than

0.01 millisievert per year from it. Compare that with the 2–3 millisieverts per year that every person receives naturally from soil, cosmic rays, and food. Put differently: Eating a banana a day, taking a medical X-ray, or flying across the country exposes you to more radiation than living beside a reactor for a year.

For additional context, a dental X-ray is about 0.005 mSv, a chest X-ray about 0.1 mSv, and a chest CT scan about 7 mSv — orders of magnitude above plant-related public exposure and still considered acceptable when medically justified.

This surprises many parents, because for decades, the public conversation has been dominated by fear. But scientists who monitor radiation exposure make the comparison clear: The everyday risks families accept without question — driving cars, breathing polluted air, crossing busy streets — are far greater than the negligible risk from living near a nuclear facility.

Long-term health studies around nuclear facilities have repeatedly failed to show population-level effects from normal operations, while the benefits of reliable, non-polluting electricity are immediate and measurable.

Why Fear Persists

If radiation is so ordinary, why are so many families afraid of it so intensely? The reasons are layered.

The Cold War left a long psychological legacy. For decades, children practiced emergency drills in schools while adults absorbed stark images of atomic tests and geopolitical brinkmanship. That era embedded a powerful association between radiation and destruction — one that continues to shape reactions long after the politics that created it have faded.

Accidents have also played a role. The names Chernobyl and Fukushima loom large, even though the circumstances behind them were specific and different from modern reactors today. Images of abandoned towns and evacuations create powerful emotional memories that are difficult to separate from everyday nuclear power.

And then there is the nature of radiation itself. Unlike smoke or smog, radiation cannot be seen, smelled, or touched. It is invisible, and the human mind fills that invisibility with dread. Fear grows where knowledge is absent.

Risk psychologists call this the "dread factor": Unfamiliar, invisible hazards feel worse than familiar ones, even when measured risk is lower.

Radiation Compared to Real Energy Risks

Perspective is essential. Families worry about radiation from nuclear plants, but far fewer worry about the pollution from nearby highways, factories, or coal plants — even though those pollutants are much more dangerous.

Air pollution from fossil fuels causes millions of premature deaths each year. It triggers asthma in children, increases heart disease in adults, and lowers life expectancy in communities exposed over decades. By contrast, normal radiation from nuclear energy has not been shown to cause health effects in surrounding populations.

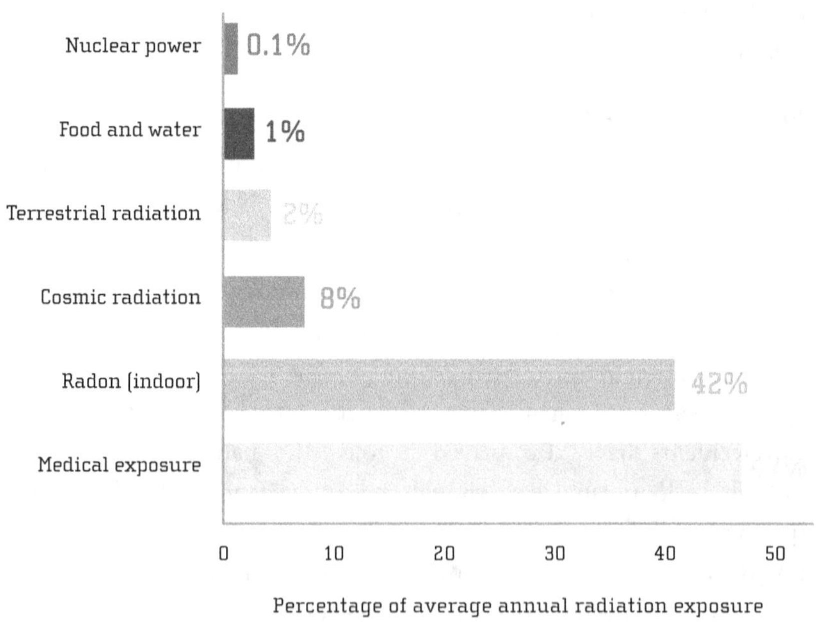

Fig. 5. Average sources of annual radiation exposure worldwide.

When researchers compare full life-cycle impacts, nuclear power's air-pollution burden is among the lowest of all major electricity sources, alongside wind and solar. The largest health gains often come from replacing coal and diesel with any non-combustion source, nuclear included.

This is not to say that accidents should be ignored, or safety dismissed. But it does mean that we need to weigh risks fairly. A child living next to a busy diesel bus depot is breathing toxins daily that are far more harmful than the trace radiation from a nuclear plant. Yet, in most conversations, the bus depot feels ordinary and unthreatening, while the plant feels terrifying.

Clearer comparisons help families focus on the risks they can meaningfully reduce today. Chief among them: dirty air from combustion.

Radon: The Risk at Home

If there is one radiation risk that families should take seriously, it is not something coming from reactors. It is radon gas seeping into basements.

Radon forms naturally as uranium in the soil breaks down. It drifts upward, and in some houses, it can accumulate in enclosed spaces. Long-term exposure to high radon levels increases the risk of lung cancer, especially in smokers.

Public-health agencies advise testing because radon accounts for a large share — often 30–50 percent — of the average person's annual radiation dose, and elevated levels vary house to house, even on the same street.

The good news is that radon is easy to test for and, if necessary, reduce. Small test kits are widely available. If levels are high, better ventilation or mitigation systems can bring them down. Unlike the abstract fear of reactors miles away, radon is a practical, manageable issue families can act on right at home.

In many countries, the action level is around 200 Bq/m^3 (about 4 pCi/L in the United States); simple fan-and-vent pipe systems typically cut levels by 50–90 percent. Bq/m^3 (becquerels per cubic metre) and pCi/L (picocuries per litre) are both units used to measure radioactivity concentration in air or water, commonly in the context of radon gas levels.

Radiation in Medicine: Trusted and Familiar

Perhaps the most compelling example of perspective is medical care. Families routinely accept radiation when it comes from doctors.

Children have X-rays when they break a bone. Parents undergo CT scans to diagnose illness. Grandparents receive radiation therapy to shrink tumours. These procedures involve doses far higher than everyday background exposure. Yet families consent without fear because the benefits are clear, and the risks are explained.

A mammogram is roughly 0.4 mSv; a head CT scan is around 2 mSv; a PET/CT scan can be 10–25 mSv. These are meaningful doses, but the diagnostic or therapeutic benefit typically outweighs the risk by a wide margin.

The medical use of radiation demonstrates a crucial truth: Radiation itself is not inherently good or bad. What matters is the dose, the purpose, and the balance of risks and benefits.

The same principle — optimize benefits, minimize dose — underpins radiation safety in power generation, industry, and research.

Replacing Fear with Understanding

Radiation's reputation has been shaped more by history, culture, and imagination than by science. For families, the challenge is not to ignore it, but to place it in context.

We live in a world where radiation is constant — from the ground beneath us, from the sun above, from the food we eat and the air we breathe. The extra radiation from nuclear energy is minuscule in comparison, dwarfed by everyday exposures we already accept. The real dangers to family health come from pollution, poverty, and fragile grids, not from the trace radiation of reactors.

Learning the simple language of dose — microsieverts and millisieverts — turns a shadowy fear into something concrete and comparable.

When we understand this perspective, the fear begins to fade. Radiation becomes less a shadow in the background and more a fact of life — one we can manage wisely, the way we manage sunlight or medical treatments. Knowledge turns dread into calm, and calm makes room for clearer choices.

Families who test for radon, wear sunscreen, and support clean electricity policies are already practicing this kind of practical wisdom.

What Families Should Remember

Radiation will always sound intimidating. But the reality is far more ordinary. Families already live with it in bananas, flights, countertops, and sunlight. They accept it in medical care. They manage it at home with radon tests.

Seen in perspective, radiation is not the great enemy it is often imagined to be. It is part of nature, part of technology, and part of life. By replacing fear with understanding, families can make energy decisions grounded in fact rather than myth. And in doing so, they can focus attention where it matters most — on building cleaner, safer, and fairer energy systems for the future.

When conversations shift from vague fear to clear comparisons — how a flight, an X-ray, or a granite counter stacks up to living near a plant — families gain confidence to choose solutions that protect both health and climate.

Family Action:

- Learn to recognize common sources of everyday radiation, including medical imaging, air travel, and natural background exposure.
- Keep a record of medical scans and ask clinicians when lower-radiation or nonradiation alternatives may be appropriate.

CHAPTER 4

Living Wisely with Energy and Radiation

Replacing fear with understanding

Understanding Instead of Avoiding

"Mom, the lights won't turn on."

It was just after seven on a January morning when Taylor's voice cut through the stillness of the Jackson house. At first, her mother, Emily, thought it was a tripped breaker — maybe the kettle had overloaded the circuit. But as she padded down the hall in her socks, she sensed the difference. No low hum from the fridge. No glow from the thermostat. No whoosh from the furnace. The house was dark and cold, and not in the usual winter-morning kind of way.

Within minutes, everyone was awake, blinking at the silence. Jacob, thirteen, tugged his hoodie tighter and mumbled, "Did we forget to pay the bill?"

Emily laughed, but her voice carried an edge. This wasn't a missed payment. This was something else. The blackout had arrived.

By mid-morning, the indoor temperature had slipped below 15°C. Taylor huddled under a quilt on the couch, flashlight in hand, like a tiny explorer. The fridge stood quiet, the internet gone, the kettle stubbornly cold. What had once been invisible — the electricity that ran their lives — had suddenly become glaringly absent. Outside, frost rimed the windows. Inside, every sound was magnified: the creak of floorboards, the faint drip of a leaky tap, the whisper of their own breath.

Across Canada, outages like this are rare but rising. Natural Resources Canada reports that extreme-weather events have already doubled the average annual hours of outage over the past decade, a trend echoed worldwide as storms intensify and heatwaves strain power lines. Families like the Jacksons are discovering what experts have long warned: The grid is a marvel, but not a guarantee.

The World Without Power

The Jacksons' blackout was unusual for Toronto, but elsewhere it is daily life. In Lagos, Nigeria, "light-off" days strike several times a week. Those who can afford generators endure the roar and the expense of burning fuel through long humid nights. For poorer families, outages mean no refrigeration, no fans in sweltering heat, and no running water when electric pumps fail.

When the power goes out for hours at a time, South African families don't just lose electricity — they rearrange life itself. Children study by candlelight, kitchens fall silent, and even hospitals and maternity wards brace for outages announced well in advance.

India's 2022 heatwaves pushed grids to the brink, prompting rolling blackouts that left hundreds of millions without fans or refrigeration during 40°C (104°F) days.

In rural Alaska, storms can isolate villages for weeks, leaving families dependent on wood stoves and community centres for warmth.

In Bangladesh's coastal delta, saltwater intrusion corrodes wiring and transformers, and every cyclone threatens another long, dark night.

Each case shows that the consequences fall hardest on households. Children miss schoolwork, parents lose wages, elderly neighbours face dangerous heat or cold. Energy insecurity is not a mere technical inconvenience — it is a human crisis. As stated earlier, roughly 770 million people worldwide still lack any access to electricity at all, a figure that shapes education, health, and economic opportunity. When power is scarce, vaccines spoil, small businesses falter, and evening study becomes impossible. Electricity, families learn, is as fundamental as clean water.

The Kitchen Table Turns into a Map

That evening, by candlelight, the Jacksons gathered at the kitchen table. Jacob pulled out a deck of cards. Taylor doodled stars and planets in the margins of her homework. Without the soft blue glow of screens, the only light came from the flicker of candles and the occasional sweep of car headlights across the wall. The blackout gave them something they didn't expect: unhurried time together, punctuated by new questions.

"I never realized how much we rely on electricity," Jacob said, laying down a queen of hearts.

"What makes our electricity, anyway?" Taylor asked.

Emily hesitated, then guessed: "Mostly hydro and natural gas here. Maybe a little wind."

The questions kept coming. How do power lines work? How long can a freezer keep food cold? What happens if outages last for days? For the first time, electricity was not just a background hum. It was a subject of curiosity and conversation.

The next night, they decided to turn that curiosity into something concrete. On a sheet of paper, they drew their home at the centre and began sketching arrows: to the furnace, the fridge, the lights, the internet. Then they traced the lines backward. Where did that power come from? A river? A nuclear plant? A gas pipeline? Soon, the kitchen table was covered in what Jacob proudly called the "Family Energy Map."

What started as frustration had turned into discovery. The blackout became the spark for the Jacksons' first step toward energy literacy, a skill many experts now argue is as vital as financial literacy in a changing climate. Around the world, educators are beginning to teach this explicitly, from "energy diaries" in German schools to community mapping projects in Kenyan villages.

Emily, remembering a neighbour's health scare, added another dimension to the map: radiation. Where does it naturally occur in daily life? She explained to the children that radiation isn't only about distant reactors. It lives in the granite bedrock beneath their feet, in the cosmic rays that streak through airplane cabins, even in the tiny sensors of their smoke detectors.

"It sounds scary," she said. "But most of it is harmless. And some of it saves lives."

Resilience Isn't Just for Engineers

That weekend, bundled against the snow, the Jacksons walked their neighbourhood. They noticed homes with solar panels under a thin white crust, and one yard with a small wind turbine. At the library, a sign announced plans for a Community Resilience Hub, promising backup power, clean-air filtration, and charging stations during emergencies.

Emily nodded as she read. "This is smart. Not everyone can afford a generator. Places like this could be lifelines."

Jacob asked if they could do anything at home. That night, back at the map, they added a new section: "What we can do." The list began modestly: Apply for the home-efficiency rebate Emily had read about; install a programmable thermostat; talk to the landlord about insulation; stock an emergency kit; check on Mrs. Peterson next door, who depended on an electric oxygen machine.

They even discussed small radiation-related steps, like testing for radon gas in the basement, a naturally occurring radioactive hazard that many Canadian families overlook. Emily recalled a neighbour who had installed a simple radon detector after a community health talk.

"Radiation isn't just reactors," she explained. "It can be right under our feet."

None of these actions alone would keep the lights on forever. But together, they built resilience. The Jacksons realized resilience wasn't just the job of engineers in control rooms. It was also a responsibility shared by families and communities. Studies show that every dollar invested in home efficiency and backup power saves several dollars in disaster-recovery costs, a reminder that preparation is often cheaper than repair.

From Puerto Rico's solar-powered microgrids to Japanese neighbourhood drills for earthquake and tsunami preparedness, families around the globe are proving that resilience begins at the kitchen table long before it reaches the control room.

A Different Kind of Inheritance

The blackout passed. Power returned at 2:13 a.m. with a chorus of humming appliances and blinking clocks. But something in the Jackson household had shifted.

Taylor began saying goodnight to her lamp as she switched it off, whispering, "Thank you, electricity," before curling under her blanket. Jacob started sketching ideas for a school project: maybe a mini wind turbine to charge phones during the next outage. Emily began keeping a closer eye on community programs and rebates, determined to make their home more efficient and secure.

The Jacksons had stumbled into something profound: energy awareness. They realized electricity wasn't just a bill or a background hum. It was a lifeline; one they could no longer take for granted.

Around the world, families who have lived through prolonged outages often report similar changes: greater conservation at home, more interest in rooftop solar or neighbourhood microgrids, and a stronger sense of shared responsibility. After the Fukushima accident, for example, Japanese households cut power use dramatically, proving that conservation on a national scale begins with tiny household habits. In rural Alaska, families teach children how to maintain generators and wood stoves as carefully as other families teach them to drive.

Families often pass down heirlooms or traditions, but energy literacy can be a different kind of inheritance, one that equips children not only to appreciate power, but to shape its future.

What Living Wisely Really Means

Living wisely with radiation — and with energy more broadly — doesn't mean never having to worry, and it doesn't mean blind trust. It means replacing fear with understanding, and ignorance with action. It means testing for radon at home rather than panicking about reactors far away. It means trusting medical professionals who use radiation for healing, while holding governments accountable for transparency in energy decisions.

Most of all, it means shifting from passive consumption to active participation. Families who know where their power comes from, who ask questions, and who take small steps toward resilience are not just customers of the grid. They are part of the energy system itself.

Simple steps — like joining a community solar project, keeping a stocked emergency kit, or learning how to read an electricity meter — can turn anxiety into agency. Even teaching children how smoke detectors use tiny bits of radiation to save lives transforms a mysterious word into everyday safety.

From Fear to Confidence

The Jacksons' blackout lasted less than twenty-four hours, but it changed their relationship with energy. They began to see electricity not as an invisible given, but as a fragile and precious resource. They learned that resilience takes shape in lived choices — how homes prepare, how communities organize, and whose voices are heard.

Across the world, families in Lagos, Puerto Rico, South Africa, and beyond are learning the same lesson, often under much harsher conditions. Outages test not only grids, but communities. The question is not whether we can eliminate all risks, but whether we can meet them with knowledge, preparation, and solidarity.

Living wisely with radiation, and with energy in general, means knowing enough to replace dread with confidence. It means raising children who understand that electricity doesn't come from magic, but from choices. And it means preparing homes and neighbourhoods not only for sunny days, but for storms and blackouts too.

Families who take these steps aren't just coping with outages. They're shaping a future in which energy is cleaner, safer, and shared more fairly. And that is a legacy worth passing down.

Family Action:

- As a family, create a simple "energy map" of your home. Sketch where your electricity comes from, how it powers daily life, and note sources of radiation in the home, including natural sources such as radon and man-made sources such as ionization smoke detectors.
- Build a small emergency kit with flashlights, batteries, and essentials, and test for radon in your basement.
- Even small habits, like turning off lights together or trying a "power-free evening," can turn energy from something invisible into something understood — and transform fear into resilience.

CHAPTER 5

How Nuclear Power Works

Splitting atoms in plain language

Making Sense of Something That Sounds Complex

For many people, the phrase "nuclear energy" immediately conjures dramatic imagery — scenes borrowed from films, documentaries, or old broadcasts — rather than the quiet mechanics of power generation. Before understanding how nuclear plants actually work, most of us first encounter the idea through imagination. For many families, the phrase feels heavy, intimidating, even a little dangerous.

But the reality is both more ordinary and more remarkable than those images suggest. At its core, nuclear energy is just another way to make heat. From that heat, we make steam. From that steam, we spin turbines. And from those turbines, we generate the electricity that runs our fridges, lights our homes, charges our children's tablets, and keeps our furnaces humming on winter nights.

If you've ever boiled a pot of water, you already understand the essence of nuclear power. The mystery isn't in the boiling, but in how the heat is made — not by burning wood, coal, or gas, but by carefully splitting the smallest building blocks of matter.

Globally, this simple idea provides roughly *10 percent of the world's electricity* and almost *a quarter of all low-carbon power*, a share that has kept millions of tonnes of carbon dioxide out of the atmosphere every year.

What Really Happens Inside a Nuclear Power Plant

At the heart of every nuclear power station lies a process called fission. The word may sound intimidating, but it simply means splitting atoms. The atoms in question are usually uranium, a naturally occurring metal found in rocks all over the world.

Here's where the scale becomes astonishing. A piece of uranium smaller than your fingernail contains as much energy as several tonnes of coal. Inside a reactor, uranium is shaped into tiny ceramic pellets. These pellets are stacked into long metal tubes called fuel rods, and dozens or even hundreds of rods are packed together in the reactor core.

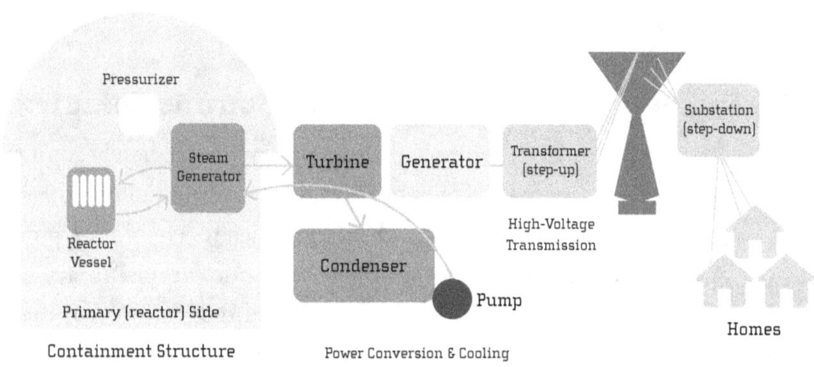

Fig. 6. Simplified pressurized water reactor (PWR), nuclear power plant, and electricity delivery pathway

When a neutron — a subatomic particle — hits a uranium atom, the atom splits into smaller atoms. This splitting releases both heat and more neutrons. Those neutrons then go on to split more uranium atoms, releasing more heat and more neutrons. A chain reaction begins.

But don't imagine an explosion. This is not a bomb. The chain reaction in a power plant is carefully tended and controlled. Think of it like a campfire. Left unchecked, a fire can roar out of control, but with steady attention, careful placement of logs, and the right balance of air, it becomes a warm, steady flame. The reactor is a campfire tended with extraordinary precision, producing a reliable supply of heat without ever letting the blaze get wild.

Modern plants use computer-guided sensors to monitor this process hundreds of times per second, keeping temperatures and pressures within safe limits.

Turning Heat into Electricity

So far, we have heat. The next step is to turn it into power.

The heat from fission is used to boil water, producing high-pressure steam. That steam surges through massive turbines — imagine giant spinning fans the size of a house. The turbines are connected to generators, which convert the spinning motion into electricity. From there, the electricity flows into the grid, travels through wires, and arrives at your light switch.

When the steam has done its job, it must be cooled back into water so the cycle can begin again. Cooling towers — those huge concrete structures often seen in photos of nuclear plants — or nearby rivers and oceans help release the excess heat. The water is condensed and pumped back into the system in a closed loop. No radioactive material escapes into the environment. The process runs over and over, quietly and efficiently, while families go about their daily lives.

To your home, the electricity from a nuclear plant looks no different than electricity from a coal plant or a solar panel. The difference lies in what's released into the air. Here, nuclear's advantage is profound: *virtually no carbon dioxide, soot, or mercury.*

Keeping Safe: How the Reaction Is Controlled

One of the first questions families ask is: If atoms are splitting and heat is building, how do we know it won't get out of control?

The answer is found in control rods. These rods, made of materials that absorb neutrons, can be inserted or withdrawn from the reactor core. Insert them deeply, and the reaction slows. Pull them out slightly, and the reaction speeds up. If anything unusual happens, the rods can drop in automatically, shutting the reaction down within seconds.

It's much like adjusting the flame on a gas stove. You can simmer gently, or you can boil briskly. And if something goes wrong, the burner can be switched off instantly.

Beyond control rods, modern nuclear plants are designed with multiple layers of protection. The reactor vessel itself is surrounded by thick concrete and steel. Cooling systems have backups for their backups. Operators are highly trained and tested, and regulators enforce some of the strictest safety standards of any industry in the world. Nuclear safety is not a single wall, but a series of walls nested inside one another.

International statistics bear this out: The *nuclear industry has one of the lowest accident-injury rates* of any major energy sector, lower even than wind and solar construction.

How Nuclear Differs from Other Power Plants

In principle, all power plants do the same thing: They boil water to make steam, which spins turbines and produces electricity. Coal burns. Gas burns. Even some solar plants use mirrors to focus sunlight and boil water. Nuclear, by contrast, doesn't burn anything at all. Instead, it taps into the binding energy inside atoms.

The result is cleaner air. A coal plant releases smoke that darkens skies and clogs lungs. A gas plant emits carbon dioxide, the main driver of climate change. A nuclear plant produces no air pollution and no greenhouse gases during operation. The only thing released into the air from a cooling tower is water vapour — steam, like that from a boiling kettle.

A single uranium fuel pellet—about the size of a small marble—contains as much energy as about one tonne of coal or nearly 400 litres of oil. That's why nuclear fuel is considered extremely energy dense, allowing vast amounts of power to be produced from very small quantities of material. For households, that difference shows up far beyond the reactor:

fewer mines carved into landscapes, fewer fuel trains crossing communities, and less pollution drifting over the places where children play.

What About the Waste?

No conversation about nuclear energy is complete without addressing waste. It is true that used nuclear fuel remains radioactive for a very long time, and that is why it must be managed carefully.

But here are the facts: Nuclear waste is small in volume, solid in form, and tightly controlled. Used fuel rods are stored in pools of water or in dry casks made of thick steel and concrete. They are monitored, catalogued, and secured. In Canada, all of the used nuclear fuel produced since the 1960s could fit into a few hockey rinks (about 60 by 30 metres), stacked a few metres deep. Compare that with fossil fuel waste, which is invisible: billions of tonnes of carbon dioxide and other pollutants released into the atmosphere every single year, where no one can contain or control them.

Countries are also planning for the very long term. Finland, for instance, is building a deep geological repository where waste can be stored safely for thousands of years, isolated far below the surface in stable rock formations. It is one of the most carefully studied and engineered projects in history, designed with future generations in mind.

Sweden and Canada have similar projects underway, and every major nuclear nation requires robust, trackable storage from the moment a fuel rod is removed.

Newer, Smarter Reactors for the Future

The reactors built in the 1970s and 1980s were milestones of engineering. But nuclear technology hasn't stood still. Today, engineers are designing Small Modular Reactors (SMRs), compact units that can be built in factories and shipped where needed.

SMRs promise added safety, lower costs, and more flexibility. They could replace diesel generators in northern communities, provide reliable power to mines and industries, or sit alongside renewable farms to balance supply. Unlike giant traditional plants, SMRs are designed to scale — to be added like building blocks depending on community needs.

Other advanced designs use molten salts or liquid metals as coolants, allowing reactors to regulate themselves through the laws of physics. These systems are like pots that cool themselves if they get too hot. They don't rely on pumps or electricity to stay safe; they rely on gravity, convection, and expansion — forces of nature that never fail.

Several prototypes, including U.S. and Canadian SMR projects, are already moving through licensing, with first units expected in the early 2030s.

For families, this new wave of reactors means cleaner air, more resilient grids, and the possibility of seeing nuclear deployed in places where it never was before.

Can Families Live Near a Nuclear Plant Safely?

The short answer is yes — and millions already do. Around the world, entire communities live alongside nuclear plants. These plants provide jobs, fund schools, and anchor local economies. Children grow up seeing cooling towers on the horizon the way other kids see smokestacks or wind turbines.

Nuclear plants are among the most heavily regulated facilities on Earth. They undergo constant monitoring, routine emergency drills, and rigorous safety inspections. Statistically, they have fewer workplace accidents than many common industries, from construction to trucking.

Long-term health studies consistently find *no increased cancer rates* for residents living near properly operated plants.

Families who live nearby rarely give the plant much thought, except when they appreciate the clean air compared to that of coal towns or when they notice the steady economic benefits.

Nuclear in the Real World

France is perhaps the best-known case. After the oil crises of the 1970s, France built dozens of reactors. Today, about 70 percent of its electricity comes from nuclear. The result: Families enjoy stable prices, a reliable grid, and some of the lowest carbon emissions in the developed world.

Ontario offers another story. By relying on nuclear, the province was able to phase out coal entirely. Smog days plummeted. Children with asthma breathed easier. Families noticed not only cleaner skies but also fewer hospital visits.

The U.S. Navy provides yet another example. For decades, its submarines and aircraft carriers have run on nuclear reactors. These ships can sail for years without refuelling. The reactors are compact, rugged, and extraordinarily safe, proving nuclear's reliability under the most demanding conditions.

Beyond Power Plants: Medicine and More

Nuclear technology also saves lives in hospitals. Radioactive isotopes are used to diagnose cancers, scan hearts, and treat tumours. Families benefit every day from nuclear medicine, often without realizing it. The same science that keeps reactors running safely helps doctors see inside the human body and target disease.

Globally, more than *forty million medical procedures each year* rely on radioisotopes derived from reactors.

In classrooms and museums, nuclear energy is becoming a teaching tool. Children explore models of atoms, build chain reaction simulations with marbles or dominoes, and learn how invisible forces shape the world around them. What was once feared can become a source of fascination, sparking the imagination of future scientists and innovators.

It's Not Magic — Just Science with Promise

At first glance, nuclear energy might look complicated. But step back, and it's really just a high-tech steam engine. Atoms split, heat is released, water boils, turbines spin, and electricity flows. The process is extraordinary in its detail, yet ordinary in its outcome: light in our homes, warmth in our rooms, and power for our communities.

Unlike fossil fuels, nuclear energy doesn't pollute the skies our children play under. Unlike intermittent renewables, it doesn't stop when the sun sets or the wind slows. Instead, it offers steady, round-the-clock power, a partner to renewables in building a clean, resilient grid.

For families, nuclear is not about abstract debates. It is about whether your child breathes clean air, whether your bills stay predictable, whether the lights stay on during the next storm. Understanding how nuclear power works turns it from a source of mystery into a source of confidence.

Family Action:

- Explain nuclear to kids in simple terms. Tell them it's like boiling water. Only instead of burning wood or gas, we use the energy hidden inside atoms.
- Remind them that nuclear plants are built with thick layers of protection, and that the electricity they help produce flows safely through the grid to power the lights in their rooms.
- Encourage curiosity. A child who learns to see energy not with fear but with fascination is a child ready to shape the future.

CHAPTER 6

The Grid — An Invisible Web

How electricity gets to your home

The Quiet Network Beneath Our Lives

Electricity feels immediate and personal: a light turning on at dusk, a screen glowing to life, a refrigerator quietly keeping food cold through the night. What rarely enters our thoughts is the vast web of wires, substations, control rooms, and power plants that makes those small moments possible. This web stretches across cities and countryside, linking distant sources of energy to everyday routines with remarkable precision. Its work is so consistent that it fades into the background of daily life, unnoticed and uncelebrated. Yet its importance becomes unmistakable when something goes wrong. A single failure can darken neighbourhoods, halt transportation, silence hospitals' backup alarms, and remind us how tightly modern life is woven together. The grid matters not because we see it, but because nearly everything we depend on rests upon its quiet, constant balance.

On a normal day, the grid is silent and dependable. Parents make coffee before work, children grab toast from the toaster, grandparents adjust the thermostat, and all of it happens without thought. But behind the walls and above the streets runs one of the most complex machines ever built — a web of wires, transformers, substations, and control rooms.

It is, in many ways, the largest man-made organism on Earth. It pulses with energy, balancing supply and demand second by second. When families in one neighbourhood switch on air conditioners during a heatwave, the grid must instantly deliver the extra power. If it falters, the consequences are immediate: Food spoils, lights dim, life-support machines fail.

For families, the grid is not just technology. It is the nervous system of modern life. Worldwide, that nervous system now stretches over *five million miles of high-voltage transmission lines* and countless local feeders — enough wire to circle the planet more than two hundred times.

A Circulatory System for Society

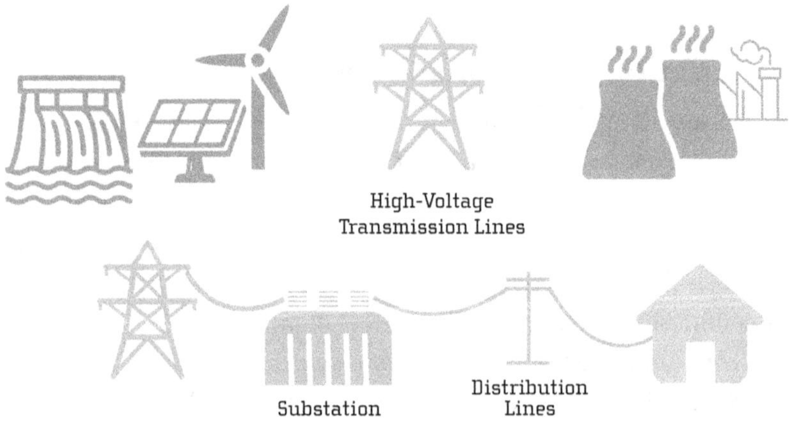

Fig. 7. The grid: How the electricity grid works

One way to picture the grid is as a circulatory system. Power plants are the heart, pumping out electricity. High-voltage transmission lines are the arteries, carrying energy across mountains, plains, and rivers. Substations step the voltage down, like lungs exchanging gases, and, finally, neighbourhood distribution lines act as capillaries, delivering power directly into homes.

The balance is delicate. Just as the human body adjusts blood flow with each step or breath, the grid adjusts electricity flow with each flick of a switch. If demand spikes — when families turn on ovens, televisions, and heaters all at once — the grid compensates. If supply dips, the balance falters, and the whole system risks collapse.

Families never see this balancing act. But they feel it the moment it fails: the silence of appliances, the stillness of fans, the sudden fragility of daily life.

Engineers call this "frequency stability," and maintaining it to within a narrow band of 50 or 60 hertz is a round-the-clock dance performed by operators and automated controls.

The electrical grid is like a circulatory system. The power plants are the heart, the transmission lines are the arteries, the substations are the lungs, and the distribution lines are the capillaries delivering energy into every home.

A Patchwork Built over Generations

The grid we rely on was not designed in a single vision. It is a patchwork quilt, stitched together over decades. Some parts are new and smart, ready to handle solar panels and electric cars. Others are aging relics, built in the 1950s, never meant to carry the digital weight of the twenty-first century.

In California, families have grown accustomed to "Public Safety Power Shutoffs," where utility companies cut power to prevent wildfires sparked by old transmission lines. Rolling blackouts disrupt dinners, cancel soccer practices, and leave children doing homework by candlelight.

In South Africa, "load-shedding" has slipped into everyday language. Families time meals and showers around blackout schedules that arrive like weather forecasts, while wealthier households hum along on generators and poorer ones sit quietly in the dark.

In Germany, the transition to renewables has strained the north-south connection. Wind farms in the north produce energy that must travel south to factories and cities. Bottlenecks in transmission mean that clean power sometimes goes unused, while families pay higher bills for imported gas.

Even in wealthy countries, the grid is fragile. In 2021, families in Texas learned this through lived experience. A severe winter storm left millions without power or heat for days. Homes grew cold. Parents melted snow for water. Children slept in coats. Some people tragically died of hypothermia. The event revealed how quickly modern life can unravel — not as a political failure, but as a human one.

North American reliability councils now warn that extreme-weather outages have *doubled in frequency over the past two decades*, highlighting how yesterday's infrastructure strains under today's climate.

The truth is that the grid was built for a simpler time: one-way power from big plants to passive consumers. But today, families are no longer just consumers. They are participants.

Families as Producers, Not Just Consumers

Rooftop solar panels are transforming family homes into tiny power plants. Batteries in basements or garages store electricity for use during blackouts. Electric vehicles double as mobile batteries, capable of feeding power back into the grid. Families are no longer at the end of the line; they are part of the web itself.

In Brooklyn, new microgrid systems are reshaping how power moves through dense urban neighbourhoods. Solar panels on one building can feed energy directly to others nearby, using digital platforms to track and settle exchanges. What once flowed only one way — from utility to customer — now circulates locally, creating a more flexible and resilient urban energy network.

In Kenya, solar-powered microgrids have electrified rural villages where the national grid never reached. Families now power schools, refrigerate vaccines, and charge phones — all from community-owned systems.

In Puerto Rico, after Hurricane Maria left the island dark for months, families and communities built microgrids to prevent future disasters.

Solar panels, coupled with batteries, now allow neighbourhoods to stand apart from the central grid when storms strike.

These stories show a shift: The grid is no longer just top-down. It is becoming a conversation between families, neighbourhoods, and nations.

Globally, distributed energy like this is projected to supply *nearly 30 percent of new electricity capacity by 2030*, turning homes into active partners in the energy system.

The Fragility of the Everyday

The grid's importance is clearest when it fails. The 2003 Northeast blackout left fifty million people across the United States and Canada without power. Families in New York cooked meals on sidewalks; rural households in Ontario spent nights in silence. Generators buzzed where they existed; elsewhere, people waited.

When the Texas freeze of 2021 hit, the sudden quiet was jarring. Heat vanished, lights went dark, and ordinary routines collapsed. Families layered clothing indoors, some burning whatever they could find for warmth. Phones were conserved like lifelines, hospitals scrambled to adapt, and entire communities were forced to confront how quickly modern life unravels when electricity disappears.

Failures expose fault lines. Some households switch on generators and continue with little disruption, while others sit through heatwaves or cold nights with no relief. For older adults and those dependent on medical devices, the stakes are far higher. In these moments, the role of power becomes painfully clear: It is a foundation of security, not a luxury.

Worldwide, the World Bank estimates that outages already cost developing economies *over $100 billion annually* in lost productivity and spoiled goods.

Smart Grids and the Future of Family Power

But change is coming. The next generation of grids will be "smart" — able to communicate, adjust, and anticipate.

Smart meters already allow families to see their energy use in real time. Parents can track when the dishwasher runs or when the heat turns on, and adjust to save money. Utility companies can send signals asking families to reduce use at peak hours, offering bill credits in return.

Vehicle-to-grid technology is also emerging. Electric cars can one day act as backup generators, feeding power back into homes during outages. A family EV might not just take kids to school. It might keep the lights on when the grid fails.

Community resilience hubs are being built in libraries, schools, and community centres. They provide safe places with backup power, clean air, and charging stations during disasters. For families, these hubs are lifelines — places where children can study, elders can breathe cool air during heatwaves, and parents can recharge phones to stay connected.

Pilot projects in Japan, California, and Europe already show that *neighbourhood microgrids and EV batteries can keep essential services running for days* after a wider outage.

The Grid and Clean Energy

The clean energy transition depends as much on wires as on wind turbines. Solar farms are often in deserts. Wind turbines spin in remote plains or offshore coasts. Hydroelectric dams sit in valleys far from cities. To reach families, this energy must travel across new transmission lines. But in many countries, those lines don't yet exist.

In the United States, thousands of renewable projects are waiting in a queue for grid connections. In Europe, bottlenecks delay clean energy from flowing freely between countries. Without grid upgrades, families end up paying more, while fossil fuels linger.

This is where nuclear, hydro, and renewables must be seen as partners. Nuclear provides steady baseload power, often near urban centres. Renewables add flexibility, often from afar. The grid is the meeting place where these sources converge. Without strong grids, even the cleanest energy cannot reach your kitchen socket.

Resilience and Equity

Grid resilience is not just a technical challenge; it is a moral one. Who gets their power back first after a storm? Which neighbourhoods wait longest? Which families can afford backup batteries, and which cannot?

Equity means designing systems that don't leave anyone behind. After Hurricane Sandy, New York invested in microgrids to protect vulnerable communities. In Canada, Indigenous-led transmission projects are connecting remote First Nations to clean power for the first time, replacing diesel with shared ownership. In South Asia, microfinance programs allow low-income families to purchase solar kits, spreading resilience to those who need it most.

Families should never be abandoned in the dark because of where they live or how much they earn. A fair grid is one that keeps everyone connected.

The Web That Holds Us Together

The grid is an invisible web, but it is also a lifeline. It keeps food cold, homes warm, schools open, and hospitals running. When it fails, the grid's silence is loudest in children's questions: Why can't we turn on the lights? When will the heat come back?

To understand energy, families must look not just at how power is made, but how it moves. The grid is not magic. It is wires, transformers, and human choices. It is fragile, but it can also be resilient. And in the years ahead, it must be made fairer, stronger, and smarter.

Families are already part of this story — through the bills they pay, the habits they form, and the rooftop panels or EV chargers they install. The future of the grid is not written in control rooms alone; it is being written in kitchens, driveways, and neighbourhoods.

Electricity is not invisible once you know where to look. It is the hum behind every bedtime story, the warmth in every winter night, the lifeline that binds us all in one great, invisible web.

Family Action:

- Next time you're together, trace the path of electricity into your home — from distant power plants, through wires and substations, to the switch on your wall. Point out nearby poles, transformers, or even rooftop solar panels, and talk about how families can also feed energy back into the grid.
- Make a simple backup plan for blackouts — flashlights, charged devices, and a neighbour to check on — so children see that resilience starts at home. By learning how the "invisible web" works, families replace mystery with awareness and take part in shaping a stronger, fairer grid.

CHAPTER 7

Families — From Consumers to Participants

How ordinary households are becoming part of the energy system

The Old Model: Pay, Use, Forget

For most of the twentieth century, families were passive energy users. You paid your bill, flipped your switch, and forgot about it. Power plants produced, transmission lines delivered, and families consumed. The relationship was one-way, like renting a hotel room: You used the service, but you had no stake in how it was built or run.

That model fit an era of centralized coal and gas stations, when electricity mostly flowed in one direction and meters were read by hand once a month. Prices barely changed from hour to hour, and few people knew (or needed to know) when power was cheapest or cleanest.

This mindset still lingers. Many families still see electricity as a mysterious utility that arrives at the outlet and vanishes with a bill at the end of the month. But a quiet revolution is underway, and families are finding themselves on the other side of the equation.

Smart meters, time-of-use rates, and phone apps now let households watch usage in real time. A parent doing laundry at 9:00 p.m. instead of 5:00 p.m. can shave dollars off a bill and help the grid avoid peak stress. In thousands of neighbourhoods, the old "pay, use, forget" story is being replaced by "plan, save, participate."

The Rooftop Revolution

Consider the Patel family in California. A few years ago, rising bills pushed them to install solar panels. At first, they thought of it as a personal savings strategy. But within months, they noticed something bigger: They weren't just consumers anymore. Their home was generating electricity for their neighbours. On sunny afternoons, their meter spun backward.

Their installer also added a simple monitoring app. The Patels began checking production at breakfast, timing dishwasher cycles for midday when their panels were strongest. A small battery let them ride through brief outages and store surplus for the evening meal.

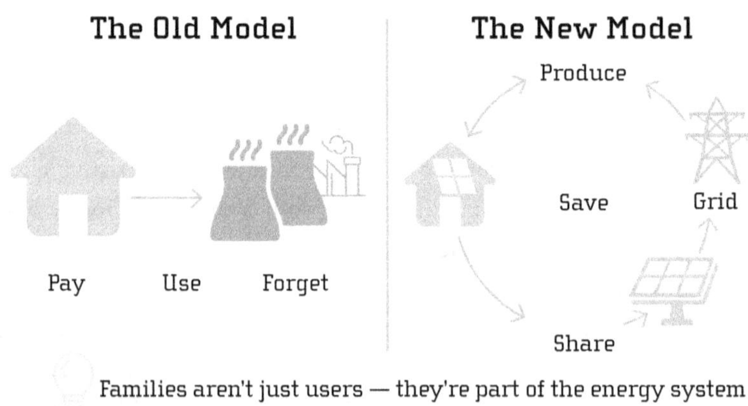

Fig. 8. From consumer to participant

Suddenly, the Patel household wasn't a passive endpoint. It was a miniature power plant. Their children bragged to friends: "Our house makes power!" Dinner conversations shifted to kilowatts and credits. Energy had moved from background noise to family identity.

Across their street, two more homes followed, pooling data to see how much of the block's usage could be met by rooftops alone. What started as one family's cost decision became a micro-movement that flattened the neighbourhood's evening peak.

Stories like this are multiplying. Across Germany, whole neighbourhoods are blanketed in solar panels, turning streets into producers. In rural India, cooperative solar farms allow farmers to irrigate fields without relying on expensive diesel. Each household that installs panels, joins a co-op, or invests in community wind shares changes the flow of energy from one-way to two-way.

In Australia, where rooftop solar adoption tops one in three homes in some suburbs, utility companies now design "solar sponge" tariffs to soak up midday generation. Families get cheaper prices when the sun is strongest, and the grid benefits too.

Community Power: Stronger Together

Not every family can put panels on a roof. Renters, apartment dwellers, or households under heavy shade are often left out. But new models are emerging that bring families together.

Shared ownership spreads costs and benefits: One project can serve hundreds of meters, and maintenance is handled professionally rather than by each household alone.

On drives beyond Halifax, the Mendes children look for a familiar stretch of open land dotted with solar panels. Their family is part of a cooperative that owns the site, and when it comes into view, the girls lean forward and say, "That's ours." What stays with them isn't the bill credit, but the feeling of belonging to something larger than their home.

The co-op also hosts weekend tours and workshops on weather-proofing and heat pumps, turning a power plant into a learning hub.

On a sunny afternoon in Brooklyn, a parent notices their rooftop panels are producing more power than their household needs. Instead of sending that surplus away to a distant utility, it goes next door — to the deli downstairs or the apartment above. In moments like this, households stop feeling like passive customers and begin to see themselves as participants in a shared energy economy.

A church in the same network added batteries to keep refrigeration and phone charging available during outages, doubling as a resilience centre for the block.

In Kenya, village cooperatives are building solar-powered microgrids that replace diesel generators. Families in these villages are not only paying less; they are part-owners of their energy future. Children in these communities are growing up with the idea that electricity is something their family can help produce, not just something bought from far away.

And when the school adds a small computer lab, the microgrid grows with it — expansion by community vote, not distant decree.

Incentives and Everyday Choices

Governments are encouraging this shift. In Canada, the Greener Homes program helps families insulate walls, install heat pumps, or put solar on rooftops. In the United States, incentives have made technologies like electric vehicles and home batteries reachable for more households, turning everyday homes into small but meaningful parts of the energy system. In Europe, feed-in tariffs once guaranteed families payment for every kilowatt they generated, sparking rooftop revolutions.

Local utility companies add their own nudges: low-interest on-bill financing for heat pumps, rebates for induction stoves, and "demand-response" programs that pay families to briefly dial down usage during heatwaves.

These incentives make it easier for families to take part, but participation also grows from everyday habits. Parents teach children to turn off lights, upgrade appliances, and run dishwashers at night. These choices might feel small, but multiplied across millions of households, they shape national demand.

A single home swapping ten incandescent bulbs for LEDs can cut roughly 400 kilowatt-hours (kWh) a year — tiny alone, huge in the aggregate. Neighbourhood challenges ("Who can cut energy use by 10 percent this month?") turn efficiency into a friendly game kids want to win.

One Ontario study showed that, if every household shifted laundry and dishwashing to off-peak hours, it would free up enough capacity to close several fossil-fuel plants. Families, acting together, can reshape entire grids.

School districts have even joined in, staggering HVAC schedules and cafeteria dishwashing to off-peak windows, saving money that flows back into classrooms.

Families as Micro-Utility companies

The energy transition is blurring lines. A household with rooftop solar, a home battery, and an EV is no longer just a consumer. It is a micro-utility. On some days, it supplies more than it takes. On others, it leans on the grid.

Add one more layer — vehicle-to-home — and the family car becomes a rolling battery that keeps the fridge cold during a storm.

Think of it like being a farmer in an energy marketplace. Some days you produce a surplus and sell it. Other days you buy what you need. The grid becomes less a one-way delivery truck and more a bustling farmers' market — with families as both buyers and sellers.

Tariffs and smart inverters now let these micro-utility companies provide tiny but valuable services — voltage support on a hot afternoon or a brief discharge to help the local feeder ride through a cloud passing over a solar farm.

This role also brings responsibility. Just as farmers must plan crops and care for soil, families must maintain systems, learn about storage, and consider resilience. The transition is not only technological but cultural: a new relationship with energy, one where families feel ownership as well as obligation.

HOA bylaws, condo boards, and landlord-tenant agreements are slowly catching up, rewriting rules to allow shared chargers, balcony solar, and battery closets that meet safety codes.

Children and the Culture of Participation

Participation is not just about kilowatts; it's about culture. Children growing up in energy-aware homes are learning that electricity is not magic but a system they are part of.

Science fairs now feature mini-grids built from Lego turbines and shoebox batteries. Kids compare output on windy versus calm days and present the findings with proud, serious faces.

In one grade five class in Kingston, Ontario, students studied their home electricity bills and debated which energy sources their province should support. One student asked: "Why don't we use more nuclear if it keeps the air clean?" Another suggested: "We should all have solar panels like my aunt." The exercise planted seeds of awareness that will shape how they vote, work, and act as adults.

A field trip to a local substation sealed the lesson: The grid wasn't abstract anymore — it had transformers you could stand next to and operators who answered questions about storms and safety.

Families can nurture this awareness by involving children in small projects — tracking household energy use, building a solar oven, or visiting a wind farm. These experiences turn abstract concepts into lived knowledge. They also create pride: Children love to say, "Our house makes energy."

Some parents set "energy captains of the week," rotating responsibility for checking the usage app and choosing an efficiency goal. Kids cheer when the family hits it.

Equity and Inclusion

But the benefits of participation are not equally shared. Low-income families, renters, and marginalized communities often face barriers: upfront costs, lack of roof access, or limited credit for financing. Without careful planning, the energy transition risks deepening inequality.

Energy burdens — the share of income spent on utility companies — can be three times higher in low-income neighbourhoods, making savings and resilience even more urgent there.

Some communities are working to bridge the gap. In Minnesota, community solar gardens specifically target low-income households, reducing bills for families who cannot install panels themselves. In Canada, Indigenous-led energy projects ensure that First Nations are not just customers but owners of clean power. In South Africa, microfinance programs allow households in townships to buy efficient cookstoves and solar kits, improving health and reducing costs.

Cities are also piloting "tariff-on-bill" retrofits so renters can benefit from efficiency upgrades with no upfront payment and guaranteed bill savings from day one.

Participation must be inclusive. A fair transition ensures that no family is left out of the chance to be a part of the future.

Practical steps — like community-owned batteries in apartment districts, EV car-share fleets, and resilience hubs in schools — turn inclusion from a slogan into something you can plug into.

From Consumers to Citizens

The shift from consumer to participant is more than technical. It is cultural, emotional, and political. Families are moving from passive users to active citizens of the energy system. They are producers, traders, advocates, and teachers.

Neighbourhood energy groups now show up at city council hearings, proposing feeder upgrades and fair interconnection rules. Parents bring kids so they can watch how local policy gets made.

This shift gives families power beyond the outlet. It gives them a voice in shaping the grid, a role in cutting emissions, and a stake in the kind of energy future their children inherit.

And the feedback loop is visible: When families invest, utility companies plan differently; when utility companies plan differently, more families can join.

Participation is not about perfection. It is about connection — families realizing that energy is not something delivered from above but something we build together. The invisible web of the grid is becoming a shared story, and households everywhere are learning to play their part.

What begins with a single rooftop or a shared solar program often ends with something larger: a community that knows its power — literally and figuratively — and uses it to light the way for others.

Family Action:

Look for one simple way your household can move from consumer to participant. That might mean shifting laundry or dishwashing to off-peak hours, joining a community solar program, or starting an "energy captain of the week" rotation with your kids to track usage. Even small steps — a smart plug, a shared solar subscription, or a family efficiency challenge — turn energy from a bill you pay into a system you shape together.

Part II

Energy Choices and Trade-Offs

Every energy source comes with trade-offs. This part looks at costs, security, health, and climate impacts of fossil fuels, renewables, and nuclear.

CHAPTER 8

Energy Costs and Security

Beyond economics — power and protection

The Quiet Weight of a Monthly Bill

Every family knows the quiet gravity of the energy bill. It might arrive by email, slip through a mail slot, or pop up as a text notification, but however it comes, it carries more than numbers. It carries the comfort of warm bedrooms and stocked refrigerators, the ability to save for a child's education or a long-awaited trip, even the simple confidence that tomorrow will unfold without sudden sacrifice.

Globally, household spending on energy now exceeds four trillion dollars each year. Economists track that figure like a barometer, but for parents, it registers in daily decisions: Another bag of groceries, or an extra sweater at night? A summer road trip, or a higher thermostat setting in winter? Tiny shifts in fuel markets — an unexpected conflict, a drought in a gas-producing region — ripple through millions of homes from Boston to Bangalore.

When the Bill Arrives

For most households, the energy story begins when the bill arrives and the family opens it together, sometimes with dread. Numbers on a page — kilowatt-hours and cubic metres — become immediate choices about food, savings, and comfort.

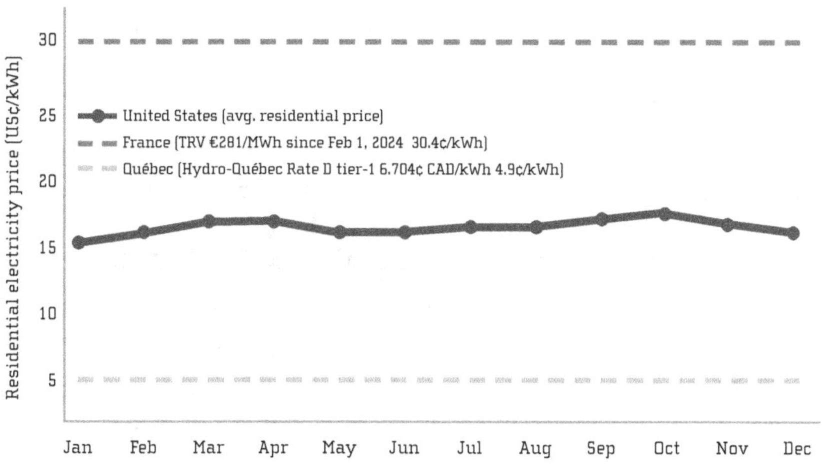

Fig. 9. Residential electricity prices in selected regions

Consider the Brown family in Manchester, England. During the winter of 2022, their monthly bill doubled almost overnight. They hadn't bought new appliances or turned up the heat. The culprit was half a continent away: global gas markets convulsed by the war in Ukraine. Suddenly, the Browns were forced into a triage of everyday life. They lowered the thermostat until their breath fogged in the hallway, cooked in big weekend batches to reduce oven use, and shelved plans for a modest holiday. Their children learned to sleep in sweaters. Anxiety threaded every conversation.

Their experience was far from unique. Across the United Kingdom, government data showed household energy costs jumping by more than 90 percent in a single year. Community centres opened "warm banks," public spaces where families could gather in heated rooms to avoid freezing flats.

Contrast that with the Nguyen family in Ontario, Canada. Their province leans heavily on nuclear and hydroelectric power. Prices nudged upward, but only modestly. Because their electricity supply was insulated from volatile global gas markets, their budget stayed predictable. They still clipped coupons and watched pennies, but they could plan ahead: a new winter coat for their youngest child, a contribution to the college fund. Stability gave them something the Browns lacked: peace of mind.

Similar contrasts played out elsewhere. In France, long-term nuclear contracts cushioned households from the wild swings that rocked gas-dependent neighbours like Germany and Italy. In parts of the United States with abundant hydropower — Washington State, for example — families rode out the same global turmoil with only minor fluctuations.

Energy Poverty: When Power Costs Too Much

"Energy poverty" sounds abstract until it lands on the dinner table. It means spending an outsized share of income on heat, light, and fuel. In the United States, the average household devotes about 3 percent of its income to energy; low-income households spend closer to 10. In remote Indigenous communities of northern Canada, where diesel remains the main fuel, some families spend more than a quarter of their income just to stay warm through eight-month winters.

Across the European Union, more than thirty-five million people are now officially classified as energy-poor, often forced to choose between heating and food. During the harsh winter of 2022, charities reported record numbers of families huddling in libraries or subway stations simply to escape the cold.

Beyond the wealthy North, the cost of power weighs even more heavily on daily life. In Lagos, Nigeria, electricity is available often enough to be counted — but rarely enough to be trusted. Households that can afford diesel generators face constant calculations, deciding when limited fuel is worth using and when darkness must be endured. A child's evening study, a hot meal, and a few hours of light all compete for the same expensive resource.

In Bangladesh's informal settlements, energy poverty takes a different form. Electricity is sold through prepaid meters that shut off the moment credit runs out. Access is immediate and unforgiving: When money disappears, so does power. Families adjust in real time, shifting homework to candlelight and watching food spoil as refrigerators fall silent. In each case, electricity exists — but only at a price many cannot reliably afford.

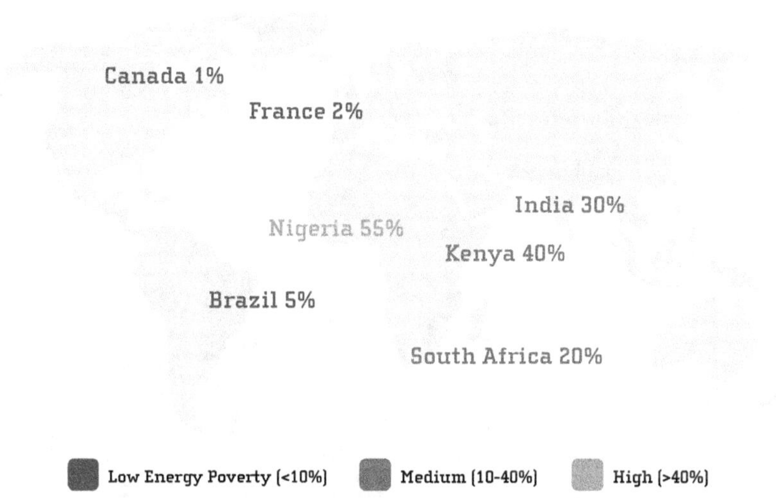

Fig. 10. Illustrative share of households experiencing energy poverty in selected countries

In Alaska and northern Canada, heating oil must be flown or barged in. A cold snap can devour a quarter of a family's monthly income in a single delivery. In the Himalayan foothills, villagers who once relied on glacial meltwater for micro-hydro power now face shrinking ice and erratic flows, making both power and water more expensive.

Energy poverty is more than a budget problem. It erodes dignity and opportunity. A child who cannot read under steady light falls behind in school. A refrigerator that cannot run endangers medicine and food. Parents who lie awake at night wondering how they'll manage the next electricity top-up aren't just juggling bills—they're carrying the heavy weight of constant uncertainty. That worry takes a toll. Research shows that living with this kind of unrelenting stress can worsen respiratory health, make it harder for children to succeed in school, and leave families more vulnerable to anxiety and depression.

The Hidden Costs of Fossil Fuels

Monthly utility statements reveal only the most visible layer of what families pay for energy. Behind every kilowatt or gallon of gasoline lies a cascade of expenses that never appear on a household budget.

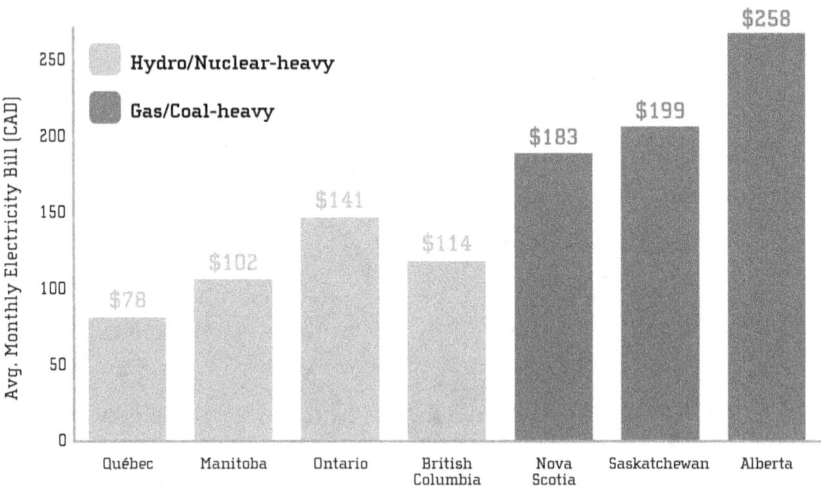

Fig. 11. Canada's illustrative monthly residential electricity costs by province (1,000 kWh/month, 2023)

When coal plants release microscopic particulates, emergency rooms fill with children suffering asthma attacks. Parents miss work to care for them, and medical bills arrive weeks later. Economists call this an "externality," but for families, it is simply a hospital co-pay or a child's laboured breathing at midnight.

Transportation fuels ripple through grocery aisles as well. When oil prices spike, the cost of shipping grain and vegetables rises, nudging up the price of bread and milk. A family might never connect a pricey loaf to a refinery outage half a world away, yet the link is direct.

Extreme weather driven by climate change adds another hidden levy. A single hurricane can destroy transmission lines and leave thousands of homes dark for weeks. Parents lose wages, food spoils, and anxiety grows.

None of these costs appear on an electric bill, yet they are borne by families everywhere.

The World Health Organization estimates that air pollution from fossil fuels causes more than four million premature deaths every year. Economists attempt to capture this burden in the "social cost of carbon," but parents translate it more personally: nights in the hospital, children kept indoors on smog days, grandparents whose lungs weaken year by year.

Security: More Than Money

Price stability is only one dimension of security. Families also need confidence that the lights will stay on when storms howl or temperatures plunge. Reliable electricity is not merely convenient; it is protection.

During the February 2021 Texas freeze, millions of households learned what insecurity feels like. Gas pipelines and poorly winterized power plants failed simultaneously. Parents melted snow on stovetops just to flush toilets. Children huddled in sleeping bags as indoor temperatures fell below freezing. Dozens of people died of hypothermia and carbon monoxide poisoning. The financial losses were staggering — nearly two hundred billion dollars — but the human toll of fear and grief was greater still.

The contrast with more diversified or robust grids was striking. In Ontario, nuclear reactors ran steadily through that same Arctic blast. Families there stayed warm. In Japan after the 2011 tsunami, inland regions with redundant transmission lines and strong disaster plans kept essential services running even as coastal towns reeled.

For parents, security is the ability to tell a child, "We'll be safe tonight," and know it is true.

Energy as Insurance

Think of dependable energy the way you think of insurance. You hope never to need it in a crisis, but when disaster strikes, you are grateful beyond measure. Reliable baseload power, neighbourhood microgrids, and home battery systems act as a silent safety net.

Across the Caribbean, communities with solar microgrids and battery storage have repeatedly restored electricity within days of hurricanes, while neighbouring islands dependent on a single central plant waited weeks. In Puerto Rico, schools that installed solar-plus-storage systems after Hurricane Maria became neighbourhood sanctuaries during later storms, keeping food chilled and medical devices running.

For families, that kind of preparedness is priceless. It is the difference between a cold night under blankets and a warm one with confidence; between a spoiled refrigerator and safe medication for a diabetic child.

Fairness and the Family Budget

Energy costs fall unevenly, and the inequalities are sharp. Wealthier households can afford to insulate attics, replace old furnaces, or install rooftop solar panels that eventually pay for themselves. Low-income families, often living in drafty apartments or older houses with outdated appliances, pay more for every degree of warmth or cool. Their bills are higher not because they use more energy, but because their homes leak it.

This pattern repeats around the world. In Germany, generous subsidies helped middle-class homeowners cover roofs with solar panels, but renters without rooftops were left behind. In India, wealthier neighbourhoods often run private diesel generators during outages, while poorer districts sit in darkness. In northern Canada, remote Indigenous communities still depend on costly diesel flown in by plane, while southern cities enjoy steady, low-carbon power from hydro and nuclear stations.

In the United States, research shows that low-income households are five times more likely to live in "energy-inefficient" homes and far more likely to suffer utility shut-offs during heatwaves or cold snaps. Those shut-offs, in turn, raise health risks, especially for children and the elderly.

For an energy transition to be truly fair, these gaps must close. Community retrofit programs, on-bill financing for insulation and efficient heat pumps, and targeted subsidies can help families lower bills while improving comfort and safety. The payoff is lasting: lower emissions, healthier homes, and smaller monthly expenses.

Stability as a Family Value

Energy costs and security are not abstract policy debates. They shape whether a parent can afford groceries, whether a child can study after dark, and whether a grandparent can breathe safely on a sweltering afternoon. A stable, secure energy system is as much a family value as honesty or kindness.

Parents in surveys consistently rank affordable, reliable electricity alongside safe schools and clean drinking water as top priorities for well-being. Without stable energy, planning for the future — college savings, home ownership, even a simple family holiday — becomes guesswork.

Looking ahead, families must demand not only affordable energy but resilient energy: power that protects as well as illuminates. Because the true cost of energy is measured not only in dollars but in dignity, safety, and the quiet confidence that tomorrow's breakfast can be cooked, tomorrow's homework finished, and tomorrow's dreams pursued without the shadow of cold or darkness.

Family Action:

Review your household energy bill together and trace where each dollar goes. Talk about what stability means for your family — comfort, safety, or savings — and explore one step to strengthen it, from sealing drafts to joining a community energy program. Security begins with awareness.

CHAPTER 9

A Future with Many Options

Blending fossil fuels, renewables, and nuclear

Choices Shape the Energy Future

Imagine sitting at your kitchen table, scanning a world-sized menu of energy choices. The spread is dazzling: sunlight captured on rooftops, atoms split inside silent reactors, rivers tumbling through turbines, and the still-familiar fuels drawn from deep underground. Families everywhere already sample from this menu without realizing it. The electricity that boils a morning kettle might be part wind, part gas, part nuclear — a hidden recipe mixed anew every second by the grid operator.

This chapter invites families to look beyond slogans or single-technology dreams and to see the bigger picture: a resilient, balanced future built from many sources of power.

The Myth of the Silver Bullet

Family life teaches that there is rarely one perfect solution. A balanced diet needs more than one food group; a family budget thrives when income comes from more than a single job. Energy is no different.

Yet public debate often slips into absolutes. Some claim solar will power everything. Others argue nuclear is the only realistic path. A few insist that fossil fuels, though flawed, are too deeply woven into society to replace. But energy researchers consistently show that the safest route is a mix. Even the International Energy Agency's most ambitious net-zero scenarios keep a variety of resources — renewables, nuclear, and some transitional fossil fuels — well into mid-century.

Balance is not a compromise. It is insurance. Just as parents diversify retirement savings to guard against market swings, societies diversify their energy portfolios to ride out heatwaves, cold snaps, or sudden price shocks. A healthy grid, like a healthy household, draws strength from variety.

Fossil Fuels: Familiar, but Waning

Coal, oil, and natural gas built the modern world. They lit the first bulbs that banished night from city streets, drove the locomotives that stitched continents together, and powered the factories that filled homes with goods. Even today, the warmth in a child's bedroom on a winter night or the gasoline that carries a family to school and work often originates in ancient carbon pulled from deep underground.

That familiarity can feel comforting. Parents know how to fill a car's tank or pay a gas bill. Many family traditions — even weekend drives or backyard barbecues — were made possible by abundant, portable fossil energy. For generations, it meant progress and prosperity.

But the other side of the ledger has always been there, even if it was easier to ignore when smokestacks meant jobs. Families inhale the cost in the air they breathe and the volatility of monthly budgets. Coal plants leave a trail of asthma inhalers and emergency-room visits. Oil's price spikes ripple through the cost of bread and milk. Gas supply disruptions — from frozen infrastructure or pipeline failures — can leave homes without heat and, at times, without electricity, plunging families into sudden darkness.

Across the globe the examples pile up. In Poland's Silesia region, miners still bring home paychecks that sustain entire towns, yet their children breathe some of Europe's dirtiest air and doctors quietly track higher rates of lung disease. In rural India, families light kerosene stoves when the grid fails, inhaling smoke that clings to walls and lungs alike. In the United States, hurricanes that disrupt Gulf Coast refineries can send gasoline prices surging within days, forcing parents to choose between filling the tank and buying school supplies.

Climate change magnifies every weakness. A heatwave in Texas can send natural-gas demand soaring and prices with it; drought in the western United States lowers river levels, straining coal shipments by barge and rail. Families feel the consequences as sudden spikes in electric bills or long waits at fuel pumps. And the warming itself — stronger storms, rising seas, longer fire seasons — adds another hidden tax to every gallon burned.

Still, fossil fuels retain powerful attractions. They are energy-dense, easy to store and ship, and supported by infrastructure built over a century. For many developing nations, they have been the fastest route to electrification. Parents in parts of Africa or Southeast Asia may see coal or gas not as relics but as lifelines that keep hospitals open and factories running.

Yet even industry leaders acknowledge the turning tide. Major oil companies now publish scenarios showing global coal demand past its peak and natural-gas growth flattening within a generation. Investment in new coal plants is drying up outside a handful of countries. Nations from Canada to China have pledged phase-downs or phase-outs. For families, this shift is both hopeful and unsettling: The old foundation is cracking, but the new one is not yet fully built.

The story of fossil fuels, then, is a story of transition. They remain the backbone of daily life for billions, but the cost — in health, climate, and economic volatility — is pushing the world toward something different. Parents explaining this to their children might say: "These fuels helped build our world, but they can't be the whole story of the future. We needed them once; now we need to move beyond them."

Renewables: Clean and Growing

If fossil fuels are the seasoned elders of the energy family, renewables are the energetic new generation — clean, fast-growing, and full of possibility. They offer something almost magical: power without smoke. When a breeze spins a turbine or sunlight pours onto a rooftop panel, no soot settles on windowsills, no carbon drifts into the atmosphere. For families, that difference is tangible: the scent of fresh air on a morning walk, the unfiltered light of a child's bedroom window, the quiet pride of knowing that electricity can be both abundant and gentle.

Across the world, households are already living this change. In Denmark, wind power has become part of the visual and cultural landscape. Turbines turn steadily above fields, coastlines, and harbours, so familiar that children sketch them without thinking twice. On especially windy days, the country produces more electricity than it can use, sending clean power beyond its borders — a quiet marker of how deeply renewables have been woven into everyday life.

In California's Central Valley, suburban streets shimmer with solar panels. Neighbourhoods compete playfully to see who can generate the largest share of their own electricity. Families check smartphone apps that track how many kilowatt-hours their rooftops have captured that day, then boast over backyard fences about shrinking utility bills. In Germany, entire villages cover barns and schools with panels, selling power back to the grid and using the proceeds for playgrounds and community centres. Clean energy, in these places, is not an abstraction; it is a local accomplishment that children grow up expecting.

Renewables are transforming daily life far beyond wealthy nations. In rural Kenya, a small solar kit can light a hut, charge a phone, and power a radio. Parents who once bought smoky kerosene for lamps now spend evenings under clear light while children read and study. In Nepal's mountain valleys, micro-hydro projects turn rushing streams into electricity for refrigeration and small businesses, allowing families to store food safely and run sewing machines or carpentry tools. In the Caribbean, rooftop solar paired with batteries keeps refrigerators running through hurricanes, giving families a sense of control when the main grid falters.

These successes inspire, but they also highlight the practical realities families must navigate. The sun sets each evening; winds can pause for days. In Germany, long winter stretches of heavy cloud cover, known as *Dunkelflaute* remind people that wind and solar output can ebb as well as flow. During these periods, turbines slow and panels produce little, highlighting a challenge that comes with relying on weather-driven power. Parents notice the subtle effects: a dimmer glow in the evening when batteries are low, or a request from the utility company to conserve until the weather shifts. Storing power for these lulls is improving rapidly, with household batteries and large grid installations becoming cheaper each year, but the challenge remains part of everyday planning.

Renewable energy offers clear environmental benefits, but it also raises questions about land, consent, and fairness. Large hydro projects can transform entire watersheds, while wind farms sometimes alter landscapes that communities depend on for livelihoods or culture. Families living near new developments may wonder why disruption arrives locally while benefits flow elsewhere. These tensions do not negate the promise of renewables, but they underline an important lesson: Clean energy succeeds best when those closest to it share in its benefits and in making decisions about its installation.

Despite those hurdles, the momentum is unmistakable. Costs for wind and solar have plunged more than 80 percent in a decade. Engineers now speak of "grid parity," the moment when clean power is as cheap as, or cheaper than, fossil fuels even without subsidies. Families sense the shift every time a neighbour installs panels or a community builds a local wind cooperative. What once seemed futuristic is becoming routine: a child plugging in an electric bike at a station powered entirely by the day's sunshine, a school boasting that its heating and lights run on geothermal steam rising silently from beneath the earth.

Perhaps most importantly, renewables change the relationship between families and energy. They invite participation. A household can become both producer and consumer — a "prosumer." Parents teach children how to monitor output and adjust usage. Communities form cooperatives where residents own a share of the local wind farm and gather annually to decide how to invest the profits. Energy, once the domain of distant utility companies, becomes something families can see, shape, and celebrate.

Nuclear: Steady and Misunderstood

If fossil fuels are the old familiar elders and renewables the eager youth, nuclear is the quiet middle child — steady, reliable, often overlooked, and frequently misunderstood. It delivers immense power without releasing carbon or smoke, operating day and night regardless of wind or weather. Compared with other energy sources, nuclear fuel packs an unusually high amount of energy into a very small volume. One small uranium pellet can replace what would otherwise require vast amounts of coal or oil, a contrast that helps explain why nuclear is often discussed differently from other power options. For families, that density translates into something tangible: refrigerators that hum through blizzards, hospital ventilators that never pause, and electric trains that glide on schedule no matter how still the air outside.

Consider Ontario, where nuclear power enabled the province to eliminate coal entirely. Parents there remember smog days when children with asthma struggled to breathe outdoors. Those days are gone. In France, where roughly 70 percent of electricity still comes from nuclear plants, families enjoy some of Europe's most stable electricity prices and one of its cleanest grids. In South Korea, nuclear plants share the landscape with bustling cities, keeping industry humming and households confident that lights will stay on even during typhoons.

Despite decades of reliable operation, nuclear energy continues to carry emotional weight. In many households, the word still evokes memories of past disasters or anxieties shaped by major accidents, creating a gap between how the technology performs and how it is perceived. News coverage of rare accidents can overshadow decades of quiet, uneventful operation. Parents may not know that modern reactors include multiple redundant safety systems or that the total volume of high-level waste produced worldwide each year would fit inside a single large warehouse.

Waste management, often portrayed as an unsolved crisis, is in fact a well-studied engineering challenge. Finland's Onkalo repository, built deep into bedrock, is designed to isolate used fuel for millennia, and the project enjoys strong local support thanks to years of open consultation. France recycles much of its spent fuel, extracting new energy and reducing volume. Engineers point out that the waste is solid, sealed, and minuscule compared with the endless stream of invisible carbon released by fossil fuels every hour.

New reactor designs strengthen the case further. Small Modular Reactors (SMRs) can be built in factories and shipped to sites, with passive safety systems that cool the core without human intervention. Some experimental reactors consume their own waste as fuel. For families in remote northern communities, these smaller plants could provide clean, round-the-clock power without the diesel shipments that currently dictate life and budgets.

Importantly, nuclear offers a partner rather than a rival to renewables. When clouds linger or winds calm, reactors provide the unblinking baseline that keeps the grid steady. Grid operators from France to Illinois quietly depend on that reliability, knowing it allows more wind and solar to flourish without fear of sudden blackouts.

To families weighing the future, nuclear may feel like the dependable neighbour — sometimes quiet, sometimes controversial, but always there when needed. Explaining it to children might sound like this: "The sun and wind give us power when they can. Nuclear gives us power no matter what. It's not perfect, but it helps keep the world clean and bright while we build even better solutions."

Why the Mix Matters

Each source carries trade-offs. Fossil fuels are reliable but dirty. Renewables are clean but intermittent. Nuclear is steady but controversial. The future is not about choosing one winner; it is about combining them wisely.

Think of a family budget. A household with one income source is vulnerable if that job disappears. A family with several income streams — wages, savings, perhaps a side business — can weather surprises. Or think of a diet: No single food provides everything a body needs. Fruits, grains, proteins, and vegetables each contribute something essential. A healthy grid is like a healthy budget or a healthy meal: balanced, diverse, and resilient.

Energy planners call this "portfolio resilience." A grid that draws from many sources can ride out droughts that cut hydro, calm weeks that stall wind, or sudden surges in gas prices — often without families even noticing.

Global Patchworks

Different countries are already stitching their own energy quilts. France leans heavily on nuclear but is adding solar and offshore wind. Germany phased out nuclear and built vast renewable farms, yet now wrestles with dependence on imported gas. Kenya combines geothermal, hydro, and solar to reduce diesel use. China expands everything at once — solar, wind, nuclear, hydro, and still some fossil fuels — building redundancy into its system.

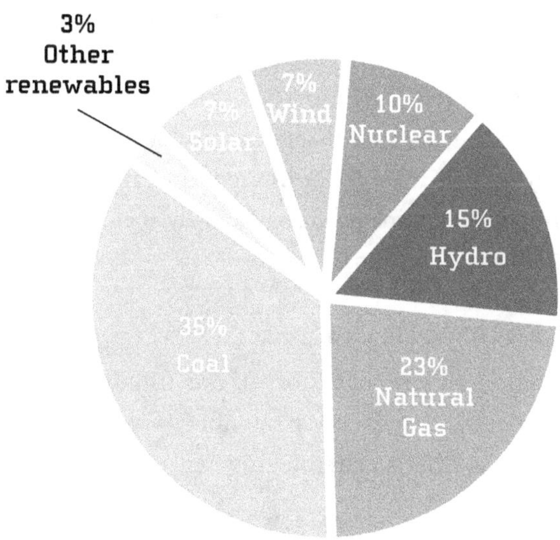

Fig. 12. Global electricity mix, 2023

Families feel these blends differently. In France, bills remain stable. In Germany, prices can spike when the wind slows or Russian gas falters. In Kenya, children can now study at night thanks to village microgrids. In China's smog-choked cities, parents press leaders for cleaner skies even as heavy industry hums.

The United States presents its own patchwork: hydro-rich Pacific Northwest, solar-heavy Southwest, wind-swept Midwest, and a nuclear-powered East Coast. Each region offers families a distinct experience of cost, reliability, and environmental impact, proving that there is no single recipe for success.

Families in the Mix

Families are not just spectators in this balancing act. They are active participants. Rooftop solar panels, electric vehicles, home batteries, and efficient appliances all shape the blend. Choices made in kitchens and garages ripple upward into policy and investment.

In Japan, families cutting energy use after Fukushima reduced demand so dramatically that the country avoided rolling blackouts. In California, households joining time-of-use programs eased strain on the grid during heatwaves. In Ontario, parents lobbying for nuclear refurbishment ensured stable power for decades. Across Africa, pay-as-you-go solar programs let households become first-time producers and customers at once, proving that participation is not limited to wealthy nations.

Every action may seem small, but collectively these decisions tilt the balance of national energy systems.

A Balanced Future

The decades ahead will not be a story of one triumphant technology. They will be a mosaic: fossil fuels fading but not gone, renewables expanding rapidly, nuclear holding steady or growing, and efficiency weaving quietly through it all.

For families, this diversity means resilience. It means fewer price shocks, cleaner air, and grids that withstand storms and crises. It means children breathing easier, parents planning budgets with confidence, and grandparents watching communities powered by many threads of energy security. Efficiency — better insulation, smarter appliances, electric heat pumps — acts like the mortar between tiles, lowering demand so every clean source stretches further.

A future with many options is not indecision; it is wisdom. Families already know the value of balance in diets, budgets, and relationships. Energy is no different. The strongest path forward is not a single road but many — converging, diverging, and ultimately carrying us together toward a stable and sustainable tomorrow.

Family Action:

- Decide your top energy priority as a household.
- Compare it with your local energy mix.
- Identify one change you would support for a better balance.

CHAPTER 10

Fossil Fuels: Familiar, but Flawed

The costs of coal, oil, and gas

The Fuels That Built the Modern World

Coal, oil, and natural gas have powered the world for more than a century. They were the engines of the industrial revolution, the spark that lit the first electric grids, the fuel that put cars in driveways and planes in the sky. Families everywhere benefited. Coal heated homes in cold winters. Oil made travel faster and cheaper. Gas stoves became a symbol of modern convenience.

Fossil fuels are energy-dense and reliable. A small amount can generate a lot of power, and the infrastructure to produce, transport, and burn them is deeply established. For generations, they felt dependable, like old recipes passed down through families: familiar, trusted, always there when needed.

But familiarity can be misleading. The very fuels that built the modern world now weigh heavily on families through health impacts, climate disruptions, and volatile costs. Fossil fuels remain widespread, but their flaws are becoming impossible to ignore.

Today, they still provide *around 80 percent of global primary energy*, but every major scientific body warns that share must fall sharply within decades to avoid dangerous warming.

Coal: The Old Workhorse with a Heavy Cost

Coal is the oldest of the fossil fuels in modern use. It drove the steam engines of the nineteenth century and still powers electricity grids in many countries today. For families in coal regions, it has long been a source of pride and paychecks.

Take a mining town in Poland. Generations of families have worked the mines, passing down not just wages but an identity tied to coal. But alongside the pride comes pollution. Children grow up breathing some of the dirtiest air in Europe, with asthma rates far above average. Parents earn livelihoods underground while their children's lungs pay the price above ground.

In China and India, coal remains the backbone of electricity supply. Families enjoy the light and warmth it provides, but smog often blankets cities, turning midday skies grey. Doctors report rising cases of chronic respiratory illness. Parents keep children indoors on "red alert" pollution days.

Satellite measurements show that coal combustion is still the single largest source of mercury in the global atmosphere, a toxin that builds up in fish and eventually in human bodies.

Coal is cheap and abundant, but it is also the dirtiest source of energy. It releases carbon dioxide, mercury, sulphur, nitrogen oxides, and particulates — a cocktail that not only fuels climate change but also seeps into families' daily lives: Instigating coughs, necessitating inhalers, and resulting in hospital visits.

Oil: Mobility and Volatility

If coal fueled factories, oil fueled mobility. It put cars in garages, buses on streets, trucks on highways, and planes in the air. For families, it opened horizons. Vacations became easier, commutes possible, and goods cheaper. Oil is woven into the fabric of daily life, from plastics to pharmaceuticals.

Yet oil carries volatility. In 1973, families in North America lined up for hours at gas stations during the oil crisis. In 2022, the war in Ukraine sent oil prices soaring, raising not just the cost of gasoline but also food and clothing, since transportation touches everything. Families felt the pinch at grocery stores, not just at the pump.

Even small supply disruptions — like a refinery fire or a hurricane in the Gulf of Mexico — can ripple through household budgets within days.

Oil dependence also creates insecurity. In Venezuela, once wealthy from oil, families now endure shortages of food and medicine after economic collapse tied to oil mismanagement. In the Middle East, families live with the dual reality of oil wealth and oil dependency: prosperity when prices are high, hardship when they fall.

For parents budgeting each month, oil is unpredictable. It provides freedom one year and anxiety the next.

Natural Gas: The "Cleaner" Fossil Fuel That Isn't Clean Enough

Natural gas is often branded as the clean fossil fuel. It burns with less soot than coal and releases fewer pollutants than oil. Families across the world rely on its bright blue flame to cook meals and heat homes. In many suburban kitchens, a gas stove is still seen as modern and desirable.

But natural gas has problems of its own. Burning it still releases carbon dioxide. Worse, leaks of methane — the main component of natural gas — trap *over 80 times more heat* in the short term than carbon dioxide. Invisible and odourless, these leaks undermine gas's reputation as climate-friendly.

Europe's dependence on imported natural gas left millions of households vulnerable when supplies were disrupted in 2022. In Germany, Italy, and the United Kingdom, rising bills forced families to stretch budgets in painful ways — turning down thermostats, wearing layers indoors, and asking children to study under blankets. The episode revealed how even so-called "cleaner" fossil fuels can carry real costs for everyday life.

Researchers estimate that unaccounted methane leaks worldwide are equal to the annual emissions of the entire aviation industry.

In many parts of Africa and South Asia, families don't have access to gas at all. Instead, they rely on kerosene or wood, which fill homes with smoke and cause deadly respiratory problems. The "cleaner" fossil fuel remains out of reach for many.

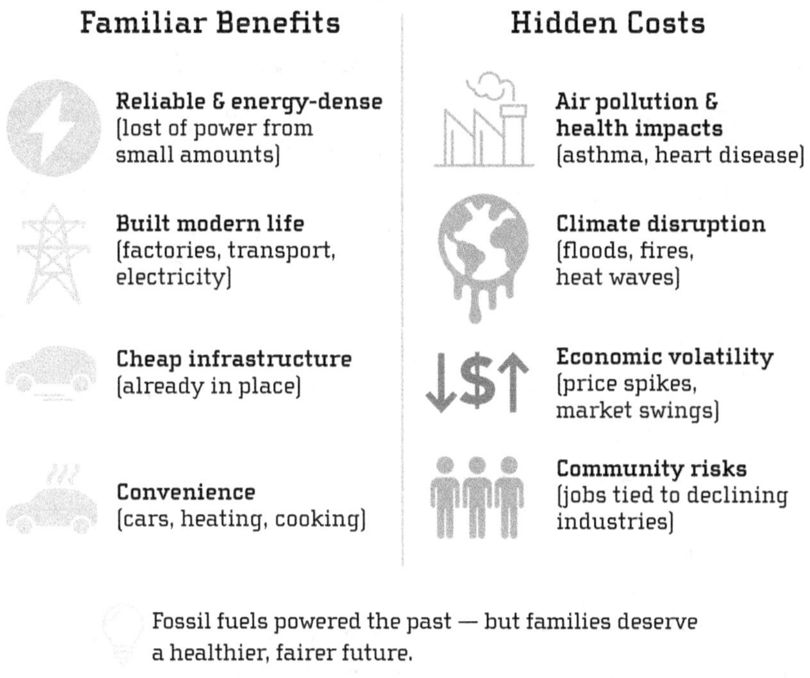

Fig. 13. Fossil fuels: Familiar benefits, hidden costs

The Hidden Health Bill

Families pay for fossil fuels not only at the pump or on utility bills, but also in hidden health costs. Air pollution from burning coal, oil, and gas kills *millions of people every year*, according to the World Health Organization. Children in coal-heavy regions grow up with lungs compromised by particulates. Parents in cities filled with traffic fumes face higher risks of heart disease.

These health impacts translate into medical bills, missed school days, and lost wages. A cheap kilowatt-hour from coal is not cheap if it puts a child in the hospital. A gallon of gas is not affordable if the smog it produces shaves years off a family member's life. Economists call these "externalities," but to families they are simply real and painful costs — hospital visits, inhalers, and shortened lives.

Climate Disruption at the Doorstep

Families are also paying the cost of climate change driven by fossil fuels. Burning coal, oil, and gas is the largest source of greenhouse gas emissions worldwide. The result is a planet of rising seas, stronger storms, and more extreme heat. In Pakistan, record floods in 2022 displaced millions of families, washing away homes and schools. In Canada and California, wildfires have sent families fleeing through smoke-choked roads, while children in distant cities breathe in haze carried on the wind. In Europe, heat domes have forced schools to close, leaving parents scrambling for childcare.

Small island nations face an existential threat. In places like Tuvalu and the Maldives, families watch the ocean creep closer to doorsteps each year, erasing ancestral land and culture.

Climate change is not abstract. It is dinner interrupted by a blackout, a child coughing through wildfire smoke, a basement filled with floodwater. These disruptions trace back to fossil fuel emissions.

Economic Volatility: Families as Hostages of the Market

Fossil fuels tie families to global markets beyond their control. A conflict in one region sparks ripples through bills in another. A cold winter in Asia can drive up natural gas prices in Europe. A refinery accident can raise gasoline costs across North America.

This volatility makes family budgeting a gamble. Parents cannot plan confidently when bills swing with wars and weather. Families live not just with financial strain, but with the anxiety of uncertainty.

By contrast, families tied to nuclear or renewable grids experience more stability. Once built, reactors and turbines don't depend on international fuel prices. Predictable bills allow families to save, invest, and breathe easier.

The Cultural Ties of Fossil Fuels

Despite their flaws, fossil fuels are deeply embedded in family cultures. Coal towns celebrate miners as heroes. Oil wealth built entire cities in the Middle East. Gas stoves became symbols of prosperity. For many families, fossil fuels are not just energy but identity.

This makes the transition difficult. Asking families to leave fossil fuels behind can feel like asking them to leave behind history, pride, and tradition. That's why the shift must be more than technical. It must be cultural — honouring the past while pointing to a healthier, fairer future.

Festivals, sports teams, and even regional cuisines carry the stamp of coal and oil wealth; transitioning means helping these communities find new sources of pride and livelihood.

Fossil Fuels as Old Recipes

Think of fossil fuels as old family recipes. They are comforting, familiar, tied to memories. But eaten daily, they are unhealthy, leaving families with ailments that grow worse over time. Just as families eventually adopt new diets for health, societies must shift to cleaner energy for survival.

Or imagine fossil fuels as a credit card. They buy us convenience now, but the bill comes due with interest — in health costs, climate disasters, and economic instability. Families are left paying debts they never signed up for.

The Challenge of Transition

Phasing out fossil fuels is not as simple as flipping a switch. Millions of families depend on fossil fuel jobs. Miners, drillers, refinery workers, truck drivers — their livelihoods are tied to industries that must decline. Communities built around coal or oil cannot be abandoned.

A just transition is essential. That means retraining workers for new industries, investing in clean energy jobs, and ensuring communities have futures as bright as their pasts. In Alberta, Canada, oil workers are being retrained to build wind turbines. In Appalachia, coal communities are experimenting with solar farms. In South Africa, programs are emerging to shift workers from coal mines to renewable projects.

Countries from Spain to Indonesia are creating "just transition funds" so pensions, schools, and local economies survive long after the last mine closes.

Families must not be forced to choose between bread on the table today and clean air tomorrow. The transition must provide both.

Familiar, but Flawed

Fossil fuels lit our homes, fueled our cars, and powered our progress. They gave families prosperity and comfort. But they also filled our skies with pollution, destabilized our bills with volatility, and disrupted our climate with consequences that reach every kitchen and classroom.

For families, fossil fuels are like that old friend who was once indispensable but now holds you back. They are familiar, but they are flawed. Gratitude for their role in history must be paired with honesty about their costs and courage to move forward.

The future will not erase fossil fuels overnight, but it will put them in their proper place: part of the past, not the backbone of the future. Families deserve cleaner air, stable bills, and a planet that feels safe to hand to their children. And that means saying thank you to fossil fuels — and goodbye.

Family Action:

- Talk together about where fossil fuels still show up in your daily life — the car you drive, the stove you cook on, or the fuel behind your electricity.
- Make a small shift as a family: Try walking or biking for short trips, using energy-efficient appliances, or choosing a "green power" option if your utility company offers it. These steps won't end fossil fuels overnight, but they help reduce pollution, save money, and show children that energy choices matter.

CHAPTER 11

Renewables — Clean and Growing, with Hurdles

Promise and challenges of wind, solar, and hydro

A New Kind of Power Rising

On a bright afternoon in California, the Johnson family's home hums with invisible energy from the rooftop. Their solar panels silently capture sunlight, turning it into electricity that powers the fridge, the Wi-Fi, and the kids' laptops. The family checks the phone app every evening and smiles: Today they produced more power than they used.

In a rural Kenyan village, a family flips on a solar lamp as evening falls. Until recently, their nights were lit by kerosene, its fumes clinging to walls and eyes. Now the air is clear, children linger over books after sunset, and weekly fuel costs quietly disappear — small changes that signal a larger shift in how power enters daily life.

Stories like these are unfolding everywhere. In the Philippine island of Bohol, the Ramos family installed a tiny wind turbine beside their coconut grove after a typhoon destroyed power lines. In northern Chile's Atacama Desert, the Cortez household helps maintain a neighbourhood battery bank that stores afternoon sunshine for evening cooking. And in Norway's Lofoten Islands, the Nygaard family heats their home with a mix of offshore wind and small hydro, finding comfort even through polar-night storms.

Globally, renewables supplied about 30 percent of electricity in 2023, a share that has doubled in just a decade. Analysts expect that figure to reach half of all power by the mid-2030s if current trends continue.

The Clean Appeal of Renewables

Renewables are clean because they don't release carbon dioxide or toxic pollutants when generating electricity. Families in cities with smog know the difference: When wind and solar replace coal, skies clear and children breathe easier.

Denmark's experience illustrates the appeal of renewable energy at scale. Wind farms there regularly meet national demand and sometimes exceed it, while in neighbouring Germany entire villages have organized themselves as energy communities, generating surplus electricity and selling it back to the grid. Together, these examples show how renewables can move from niche projects to mainstream infrastructure.

Costs have plummeted. Solar panel prices have dropped by more than 80 percent in the last decade; wind turbines now compete with natural gas on price. For families, this means rooftop solar is no longer a luxury but a realistic investment. In Australia, more than one in three homes now sports panels, while in Brazil's sun-drenched northeast, neighbourhood "solar condos" let renters share big communal arrays.

Even massive, utility-scale projects carry a sense of local pride. Morocco's Noor complex — visible from space — employs thousands of locals and lights homes across North Africa. In the United Arab Emirates, desert solar farms now deliver electricity cheaper than any fossil plant ever built.

Families notice the difference not just in their lungs or wallets but in their neighbourhoods. Children bike under clearer skies; grandparents linger on porches without the haze of coal smoke. Clean power is not an abstract policy — it is a lived daily improvement.

The Hurdles: Intermittency and Storage

Yet renewables face serious challenges, the largest being intermittency. The sun sets. The wind calms. Rivers run low during drought.

California's "duck curve" shows how solar generation peaks at noon and crashes at sunset just as families turn on stoves and televisions. When *Dunkelflaute* sets in for days or even weeks, Germany's energy system must adapt quickly. With weak sun and still air, utility companies turn to imports or restart coal plants to keep homes heated and lights on — a reminder that without sufficient storage or backup, intermittent renewables can strain even well-developed grids. Even Kenya's geothermal-rich grid must plan for dry seasons when hydro reservoirs drop.

Storage is the obvious answer, but current batteries remain expensive and resource-intensive. Mining lithium and cobalt raises concerns about water use, pollution, and labour conditions. Parents who enjoy clean skies may not see the scars left in mining towns half a world away.

Engineers are experimenting with new solutions: sodium-ion batteries made from abundant salt, iron-air batteries that store power for days, pumped-hydro reservoirs that use mountain lakes as natural batteries, even underground compressed-air caverns. All are promising but not yet widespread.

Hydroelectric dams provide steadier renewable power, but they too can harm ecosystems and communities. The vast Belo Monte dam in Brazil displaced entire river villages. Even "run-of-river" projects require careful planning to protect fish and water rights. Downstream families feel these choices when fewer fish reach their nets and when changing water levels make paddling to school, work, or the market more difficult and sometimes unsafe.

Global Patchworks of Renewable Power

Across the planet, every country is stitching its own quilt of renewable energy, and the patterns are as varied as the landscapes and cultures they emerge from. In China, entire deserts have been transformed into shimmering seas of photovoltaic panels. From the air, the solar farms of Qinghai and Xinjiang look like vast blue lakes, absorbing sunlight where once there was only sand. Families living in the nearby towns talk about the quiet pride they feel when their evening lights are powered by the same sun that bakes the desert floor. The electricity is cheaper and the air is clearer, but the view outside their windows has changed forever, reminding them that even clean power leaves its mark.

Far to the south, India is taking a different path. Instead of concentrating solely on giant utility projects, it is weaving together thousands of village-scale microgrids. In the state of Bihar, children help their parents maintain neighbourhood solar arrays, learning to tighten bolts and check wiring as naturally as other children learn to cook or farm. For households that once depended on smoky kerosene lamps, the glow of an LED bulb after sunset feels like a quiet revolution. Parents describe the sense of dignity that comes when their children can study late into the night or when fans hum softly through sweltering evenings.

In Kenya, the earth itself provides the energy. Along the Great Rift Valley, geothermal wells tap the planet's natural heat, sending steady electricity into homes from Mombasa to Nairobi. For families there, geothermal power is more than a technology; it is a dependable neighbour. Farmers who once worried about blackouts during harvest season now irrigate with confidence, and children know that the "steam that never sleeps," as locals call it, will keep their schools and clinics running even when the wind dies or the sun hides behind clouds.

Norway offers another variation on the theme. Mountain rivers and fjords supply small hydro plants, while offshore winds whip across the North Sea to giant turbines visible from fishing villages. In the Lofoten Islands, families who have depended on cod for generations now watch sleek blades turning beyond the horizon. The whir of turbines has joined the cry of seabirds and the crash of waves as part of the region's soundtrack. Residents speak of mixed feelings — gratitude for reliable, carbon-free heat during long Arctic winters, but also nostalgia for the uncluttered seascapes of their childhoods.

The United States contains several different energy patchworks within a single country. In Texas, ranchers lease their land to wind developers, turning pastures into power plants and earning enough in royalties to send children to college. On the Pacific Coast, tidal experiments share the shoreline with old hydro dams, while in the sunny Southwest, sprawling solar farms shimmer like mirages. Each region tells its own story: a farmer in Iowa counting extra income from turbines, a Navajo family debating the siting of new solar arrays, a Californian household combining rooftop panels with a backyard battery to ride out wildfire-related outages.

These examples reveal a common truth. There is no single recipe for renewable energy, no universal sequence of technologies that every nation must follow. Geography, politics, culture, and economics shape each blend. Families feel those choices in intimate ways: in the cost of a monthly bill, the colour of the evening sky, the hum of a new turbine, or the steady glow of a lamp that no longer depends on diesel. Whether it is a child in rural India doing homework under a solar light, a Norwegian fisherman warming his house with hydro power, or a Texas rancher watching turbines spin across the prairie, each household lives inside its own patch of the global renewable quilt. Together, these diverse patterns create something larger than any single project — a worldwide tapestry of clean energy that is already reshaping how families experience light, heat, and hope.

Policy and Participation: Families as Partners

Technology alone cannot guarantee success. Policies and participation matter.

In South Korea, neighbourhood cooperatives vote on where to place community solar gardens, giving residents a literal stake in the outcome. In Germany, national "feed-in tariffs" let families sell rooftop power back to the grid, turning households into micro-utility companies. In the United States, school districts from California to New Jersey are installing solar canopies over playgrounds, teaching children and saving money at once.

Indigenous leadership is equally vital. In Canada, First Nations co-own major wind projects and insist on wildlife protections and revenue-sharing. In New Zealand, Māori guardianship ensures geothermal plants honour sacred landscapes. These examples show that when families are partners — financially, culturally, and politically — renewables grow faster and conflicts fade.

The Family Dimension

Families feel the promise of renewables most directly when bills drop and air clears. Parents notice fewer smog days. Children breathe easier. Rooftop solar becomes a source of pride, a way to teach kids that the home itself contributes to a cleaner planet.

But families also encounter the hurdles. In rural Ontario, parents debate whether new wind farms spoil beloved views or lower property values. In California, utility companies ask households to shift usage to off-peak hours, testing habits built over decades. In Nairobi's informal settlements, families cheer solar lamps yet still dream of full grid service for refrigerators and computers.

Participation is key. When residents share ownership — whether through co-ops, neighbourhood batteries, or school microgrids — acceptance soars. Conflicts decline when families see their fingerprints on the project.

Renewables as Crops in the Energy Garden

One way to understand renewables is to think of them as crops in a family garden. Sun and wind are seasonal gifts. They provide abundance when conditions are right, but less in lean times. Families plan accordingly, storing food for winter or planting multiple crops to ensure a steady diet.

Renewables are similar. They can provide abundance, but only as part of a mix. Batteries are like pantries, storing energy for when it's needed. Nuclear and hydro are like staple grains, steady and reliable. Fossil fuels, though flawed, still act as emergency rations in some places. The healthiest "energy diet" is diverse — what planners call "energy horticulture."

Looking Ahead: Families in a Renewable Future

Analysts forecast that global solar capacity could triple by 2030, while offshore wind may grow fivefold, reshaping coastlines and job markets alike. Training programs are already helping former coal workers move into solar and wind careers, ensuring that the next generation of energy workers comes from the same communities that once mined or drilled.

For families, the renewable future will feel visible: more panels on rooftops, more wind turbines on hills, more conversations about energy at the kitchen table. Children who once drew coal trains now sketch spinning blades and shining panels in school art projects.

But the challenge remains: to build systems that are reliable, fair, and sustainable from start to finish. That means investing in storage, building smarter grids, and ensuring that families in mining regions are not sacrificed for families with solar rooftops. It means making renewables not just clean, but just.

Clean and Growing, with Hurdles

Renewables are clean. They are growing fast. They empower families and inspire children. But they also come with hurdles: intermittency, resource demands, land conflicts, and the need for storage.

For families, the lesson is balance. Renewables are vital, but they are not perfect. They shine brightest when paired with steady partners like nuclear and hydro, and when backed by fair policies that include everyone.

The renewable future is not a dream — it is already here, on rooftops and in villages around the world. The task now is to grow it wisely so that families everywhere can share in its promise without carrying its burdens unfairly.

Family Action:

Families can play an active role in the renewable transition by installing rooftop solar or joining community energy projects, shifting usage to sunny or windy hours, and teaching children how energy is produced and stored. By conserving power, supporting fair local initiatives, and sharing in global stories of clean energy, households help build a cleaner, more resilient future.

Part III

The Role of Nuclear in a Changing World

Nuclear power is viewed differently around the world. This part explores its role in the climate emergency, and global case studies are used to show how nuclear plays a role in justice, ethics, and public trust.

CHAPTER 12

Nuclear and the Climate Emergency

A tiny pellet with giant power

Small Sources, Far-Reaching Impact

When people talk about the climate crisis, images of wildfires and floods come to mind. For years, climate change was framed as something far away. We saw images of melting glaciers, polar bears drifting on shrinking ice floes, and maps of distant islands threatened by rising seas. Families in temperate regions could convince themselves that the problem was serious but remote — something their children or grandchildren might face decades from now.

Today, even families who never follow climate headlines can sense the change in familiar ways. Seasons feel unreliable: late snow one year, smoke-darkened skies the next, summer nights that never seem to cool. School calendars now include "heat days" alongside snow days, and parents keep air filters tucked away like earlier generations kept candles for storms. The signals are subtle at first, but increasingly hard to ignore.

That illusion has shattered. Today, families experience the climate emergency at their own front doors. Summers bring deadly heat domes, forcing schools to close and straining hospitals with children and seniors suffering from heatstroke. Wildfire smoke drifts across entire continents, turning skies orange and filling lungs with haze. Floods wash through basements, storms rip roofs from homes, and groceries cost more when droughts wither crops.

Even small changes feel big at home: the toddler who coughs during smoke season, the teen whose summer job is cancelled during a heat advisory, the grandparent who needs a cooler room and a reliable outlet for a medical device.

Ask a child in California why their soccer game was cancelled, or a teenager in Pakistan why their home was under water, or a parent in Germany why winter heating bills spiked after imported gas prices soared. The answer traces back to the climate crisis. Families no longer need to imagine what climate change feels like. They are living it.

In kitchen-table terms, climate change has moved from a far-off headline to a line item in the family budget and a note on the family calendar.

The challenge is enormous: Scientists warn we must reach net-zero emissions by mid-century to avoid the worst impacts. That means balancing the carbon we release with what we capture or eliminate. It also means transforming the way we power our homes, schools, and communities. Renewables like wind and solar are essential, but as families are learning, they cannot do the job alone. The climate emergency demands reliable, clean power around the clock. And this is where nuclear energy enters the picture — not as a rival, but as a lifeline.

Think of the clean energy puzzle as a jigsaw on the coffee table: solar and wind are bright corner pieces, efficiency is the border, storage fills in swaths, and nuclear is the piece that clicks the middle together so the picture holds.

A Tiny Pellet with Giant Power

Imagine holding a pellet of uranium in your hand. It is about the size of the tip of your pinky — small enough to lose between couch cushions.

It looks unremarkable, like a pebble from a driveway. But that pellet contains astonishing potential: as much energy as a tonne of coal, 150 gallons of oil, or 17,000 cubic feet of natural gas. Enough to power an average family home for months.

That dense energy means fewer trains of fuel rumbling past neighbourhoods, fewer ships crossing oceans, and far less land used for energy production overall.

For parents explaining this to their children, the analogy is simple: It's like packing an entire trainload of coal into a single pebble. Nuclear energy harnesses the forces inside atoms, releasing them in controlled reactions that create heat, boil water, spin turbines, and generate electricity.

In tour videos from power plants, kids often recognize the heart of the system instantly: "Oh — it's a giant tea kettle!" They're right. The science is complex, but the outcome is familiar: Steam turns a turbine, and the lights come on.

The miracle is not just the power density, but the cleanliness. A nuclear reactor produces this electricity without releasing carbon dioxide into the atmosphere. No soot. No smog. No greenhouse gases. While coal and gas plants fill the skies with pollution that worsens asthma and drives global warming, nuclear plants provide steady, large-scale electricity with virtually zero emissions.

Over a plant's lifetime, the total "cradle-to-grave" emissions — including construction and fuel — are comparable to wind and lower than almost any fossil-fueled option.

In a climate emergency, where every tonne of carbon avoided is a gift to future generations, this difference is profound.

It's the difference between a child's inhaler used daily and a child who forgets where the inhaler even is.

Reliable in Any Weather

Families know the rhythm of weather: sunny days that power solar panels, breezy days that spin turbines, cloudy or still days when renewable output falls. Renewable energy is vital, but it is also variable.

Grid operators plan for this dance hour by hour, but winter mornings, heatwaves, and long cloudy stretches still test the system.

Picture a family home on a winter morning. Snow covers the roof, solar panels are buried, and the sky hangs heavy with clouds. Outside, the air is still. Inside, breakfast is cooking, children are getting ready for school, and heaters maintain a steady warmth. The morning unfolds smoothly not by chance, but because energy arrives reliably, supporting each small, ordinary moment.

For the family running a medical device at night, or keeping baby formula refrigerated, "necessity" is literal.

This is where nuclear energy stands out. Unlike wind or solar, nuclear does not depend on weather. It works day and night, rain or shine, summer or winter. It is like a sturdy winter coat that keeps families warm no matter the storm. Parents do not have to wonder whether the wind will blow or the sun will shine; nuclear provides steady electricity, reliably, without interruption.

Nuclear plants typically run most hours of the year — a consistency planners call "high capacity factor" — which makes them ideal teammates for weather-dependent renewables.

The Texas freeze of 2021 showed how quickly energy systems can fail when extreme weather hits. As temperatures dropped, millions lost power and heat, forcing families to improvise — melting snow for water, huddling together in unheated homes. The crisis revealed a simple truth: In a changing climate, reliable electricity isn't a luxury, but something families depend on most when conditions are at their worst. Meanwhile, in Ontario, nuclear reactors kept running steadily, providing the backbone of a grid that endured the storm.

Different regions faced different problems — frozen gas infrastructure here, iced turbines there — but one pattern stood out: Systems anchored by robust all-weather generation rode through the crisis better.

Reliability is not abstract. For families, it can mean the difference between life and death.

It's the quiet certainty that the thermostat will click on, the phone will charge, and the oxygen machine will hum.

Safety First: Lessons from the Past

Of course, many parents hear the word "nuclear" and feel a twinge of fear. They remember images of Fukushima in 2011 or Chernobyl in 1986. These accidents loom large in collective memory, shaping public perception.

Memories are sticky — especially the frightening ones — and they can overshadow decades of normal, uneventful operation.

It is natural to worry. Parents worry when their children climb trees, cross busy streets, or ride bicycles. Nuclear energy feels like a big tree, a wide street — the risks loom larger than everyday dangers.

Good risk literacy starts by comparing like with like: everyday exposures, long-term health outcomes, and actual safety records across energy sources.

But it is important to understand how much has changed. Modern nuclear plants are built with multiple layers of protection — containment domes, redundant cooling systems, and passive safety features that use physics itself (like gravity and convection) to keep reactors stable even if human systems fail. Engineers describe reactors today as "treasure chests within treasure chests," each layer designed to contain risk.

Regulators now require frequent stress tests, multi-day blackout drills, and thicker physical barriers than earlier generations ever had.

One way to explain it to children is through helmets and pads: "When you ride your bike, you wear protection. Nuclear plants now wear helmets, knee pads, and elbow pads — all at once."

And they practice — constantly — so that those pads aren't just for show.

Meanwhile, families must compare risks fairly. Fossil fuels cause millions of premature deaths each year from air pollution — asthma, heart disease, and cancer. By contrast, the total global death toll from nuclear accidents, though tragic, is far smaller. Statistically, nuclear energy is among the safest sources of electricity ever used, far safer than coal, oil, or gas.

That doesn't erase grief from past accidents. It simply keeps perspective so we make choices that save the most lives.

Nuclear is not about pretending risks don't exist. It is about managing them wisely — the way families do every day in countless other contexts.

Seat belts, smoke alarms, lifeguards — all are acknowledgments that safety is something we build, not something we wish for.

The Waste Question: Teaching with Perspective

When families talk about nuclear, the word "waste" often comes up. Children imagine glowing barrels. Parents worry about long-term risks.

Curiosity is healthy — and a chance to teach proportion and planning.

Here's the perspective: All of the nuclear waste produced in the United States since the 1950s could fit on a single football field, stacked less than ten yards high. By contrast, fossil fuels release billions of tonnes of invisible waste — carbon dioxide — into the atmosphere every year, with no containment at all.

The nuclear waste volume is small because the fuel is so energy-dense; a little goes a very long way.

Nuclear waste is real, and it requires careful handling. But it is small in volume, solid in form, and tightly managed. Used fuel rods are stored in water pools or dry casks made of concrete and steel. Countries like Finland are building deep geological repositories, designed like time capsules to safely contain waste for thousands of years.

Some advanced reactor designs may even reuse portions of existing spent fuel, extracting more energy and shrinking the waste challenge further.

Parents can explain it this way: "When we make campfires, we have ash to clean up. Nuclear makes waste too, but it's kept in safe containers, not released into the air. It's like putting the ash in a sealed jar instead of letting it blow into everyone's lungs."

Kids get this quickly — containment is a concept they practice every day with lunchboxes and lids.

Counting the Costs

Another common question: If nuclear is so useful, why isn't everyone building more reactors?

A big part of the answer comes down to cost and time. Building a large nuclear plant takes billions of dollars, years of planning, and a lot of regulatory checks along the way. Delays happen, budgets creep up, and the whole process can feel slow. It's a bit like deciding to build a brick house with a steel roof—strong, reliable, and built to last, but definitely not cheap or quick upfront.

Financing costs can loom as large as concrete and steel — interest accumulates during long builds, and that shows up in the final price.

But change is coming. Small Modular Reactors (SMRs) are emerging as the "tiny houses" of nuclear power. Factory-built, transportable, and standardized, they can be deployed faster and at lower cost. For families, this means nuclear power could soon arrive in more places — from remote northern towns that now rely on diesel, to industrial hubs that need steady power.

Standard designs also let regulators review once and replicate many times — the same way commercial airplanes are certified by model, not by each individual plane.

SMRs won't replace large reactors overnight, but they represent a new chapter. They are designed with modern safety features, flexibility, and affordability in mind — the kind of innovation families expect in a twenty-first-century solution.

Think of them as neighbourhood anchors that pair naturally with wind, solar, and storage — steady hands on the grid's steering wheel.

Global Lessons: France, Ontario, and Beyond

To see nuclear in action, look to France. After the oil crises of the 1970s, France invested heavily in nuclear. Today, about 70 percent of its electricity comes from reactors. French families enjoy stable bills, clean air, and one of the lowest per-capita carbon footprints in the developed world. France even exports electricity to neighbours, helping lower emissions across Europe.

Uniform reactor designs there also created a skilled workforce and predictable maintenance cycles — a practical lesson for countries considering scale.

Or look to Ontario, Canada. By refurbishing and expanding its nuclear fleet, Ontario phased out coal entirely. Smog days vanished.

Hospital visits for respiratory illness fell. Families breathed easier — literally. Nuclear made clean air not just a climate goal, but a daily health benefit.

The province's reactors also served as the dependable base that allowed wind and solar to expand without risking blackouts.

Even countries with complicated histories are rethinking nuclear. In Japan, where Fukushima cast a long shadow, the urgency of climate change and the pain of high fuel imports are shifting perspectives. In Germany, the decision to phase out nuclear left the country reliant on coal and gas, sparking debate about whether the exit was premature. In South Korea and China, nuclear is expanding as part of climate and energy security strategies.

Different paths, same lesson: Durable, clean, always-on power is hard to replace — and families notice when it's gone.

Each story shows nuclear not as a perfect solution, but as a powerful tool. Families in these countries live the results in their bills, their air quality, and their resilience during crises.

When the grid bends but doesn't break, the difference is felt not in headlines, but in quiet, ordinary evenings at home.

Families and the Climate Emergency

For families, the climate crisis can feel overwhelming. Parents see children's anxiety about the future. Kids ask why summers are hotter, why smoke fills the air, and why leaders don't act faster.

Those questions are heavy — and they deserve answers grounded in action, not just worry.

Nuclear offers families a story of action. It shows that science can provide solutions, that societies can use powerful tools responsibly, and that no option is off the table when the stakes are survival.

It also invites participation: school visits to local plants, science projects about chain reactions, neighbourhood discussions about grid reliability and fairness.

Explaining nuclear to children is easier than it seems: "It's like boiling water, but instead of burning coal to create heat, we split tiny atoms. That way, the sky stays clear." Framing nuclear as part of a broader mix — alongside solar and wind — helps children see it not as a scary exception, but as a partner in building a livable future.

When kids help track home energy use — noting sunny-day solar, windy-day turbines, and the steady baseline — the idea of teamwork among energy sources becomes intuitive.

A Lifeline, Not a Luxury

The climate emergency is no longer distant. It is in the air families breathe, the prices they pay, the storms they endure. Every tonne of carbon avoided is a gift to children's futures. Nuclear energy, with its dense power, steady reliability, and near-zero emissions, is one of the strongest tools available.

It is the dependable friend in a crowded tool shed: not flashy, but there when you need heavy lifting done.

It is not flawless. It requires vigilance, investment, and honest dialogue. But in a world on fire, nuclear is not a luxury — it is a lifeline.

Used alongside renewables, efficiency, and storage, it can help stitch together a grid that is both clean and calm — the kind of system families can trust through smoke seasons and storm seasons alike.

For families, that lifeline means cleaner skies, steadier bills, and a safer future for their children. The question is not whether nuclear is perfect. It is whether we will act boldly enough in this emergency to give families every possible chance at stability and hope.

The clock is ticking — but turbines are spinning, reactors are humming, and children are still dreaming toward a sky that can stay bright and safe for them.

Family Action:

Frame family climate talks around shared values, near-term actions, and the role of clean, reliable power in protecting health and budgets.

CHAPTER 13

Global Perspectives — Six Countries, Six Nuclear Stories

Nuclear power doesn't mean
the same thing everywhere . . .

Germany: From Embrace to Exit — and Maybe Back Again?

In a suburb outside Munich, the Schneider family sits down to dinner as evening news murmurs in the background. Their teenage daughter, Lena, has been learning about climate change at school. Midway through the meal, she asks why Germany no longer uses nuclear power if it doesn't emit carbon. Her father pauses before answering. Germany once relied heavily on nuclear, he explains, but after Fukushima, public fear reshaped the country's energy path. The decision still lingers, not as a closed chapter, but as a question many families quietly revisit.

Around the table, they compare notes from different generations: Oma remembers anti-nuclear marches and the Cold War; Lena's class talks about smoke-filled summers and climate targets. The same word, "safety," lands differently depending on age and experience.

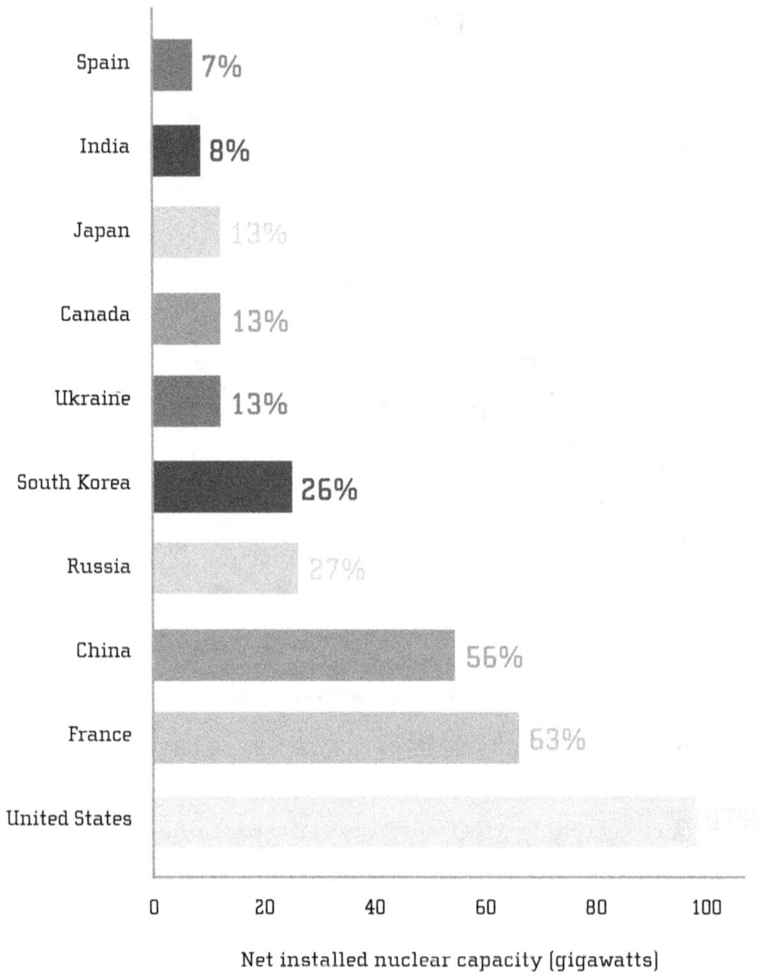

Fig. 14. Nuclear power capacity by country (top 10, 2024): net installed capacity of reactors in operation

Germany's relationship with nuclear power has been anything but linear. In the decades after World War II, the country invested heavily in nuclear energy. They built reactors across the landscape to provide reliable electricity and lessen reliance on coal and imported fuels. That trajectory shifted sharply in 2011. When the Fukushima disaster unfolded, it reverberated deeply through German politics and public opinion. Chancellor Angela Merkel announced a rapid timetable to close all remaining nuclear plants, bringing an era of nuclear generation to an official end by 2022. The policy decision was settled, but the questions it raised were not.

Utility companies planned decommissioning. Engineering programs shifted focus. Households were told the future would be built on wind, solar, and efficiency. For a time, that vision felt unstoppable.

Many families felt genuine relief. Germany's anti-nuclear movement had deep roots, shaped by decades of environmental activism and memories of technological catastrophe. For parents who had marched against reactors in the 1980s, the phase-out felt like vindication. Wind and solar expanded rapidly, and the *Energiewende* became a source of national pride. Rooftop panels appeared on barns, schools, and suburban homes, tangible proof that progress could feel both clean and safe.

Local cooperatives flourished; schoolchildren toured wind farms and learned to read home energy monitors. Posters proclaimed, "The future is ours to build," and for a while, many believed it.

But the transition was not seamless. When the sun failed to shine or the wind slowed, coal plants filled the gap. Lignite, the dirtiest form of coal, returned in force. Imported gas — much of it from Russia — became a crutch. When the war in Ukraine disrupted those imports in 2022, German families saw their bills soar. Parents turned down thermostats, and headlines warned of energy rationing.

In some neighbourhoods, "thermal socks and sweaters" became a winter uniform; bakeries shortened hours to cut electricity costs; town halls opened warm rooms for seniors.

Public opinion shifted. A 2025 survey found that, while most Germans had supported the phase-out in 2011, many now questioned whether it had been wise. Younger generations, facing both high bills and climate anxiety, began to ask why clean nuclear had been sacrificed while coal endured.

University debates grew spirited: "risk now" (coal and gas) versus "risk imagined" (nuclear accidents). Families noticed the trade-off not as theory but as receipts and air quality reports.

The debate is evolving. Some politicians hint at rethinking the ban. Families remain divided. For some, rejecting nuclear is a matter of principle — a commitment to never risk another Fukushima. For others, the urgency of the climate crisis makes reconsidering nuclear not only practical but ethical.

Town halls brim with questions about Small Modular Reactors, recycling of fuel, and what a "minimum nuclear" path might look like alongside renewables.

Germany's story shows how energy is not only technical but moral. Families weigh safety, sustainability, and independence differently, and their choices ripple far beyond their borders.

Whatever path Germany chooses next will teach the world — again — how values and volts intertwine.

France: A Nuclear Identity

Along the Loire River near the town of Chinon, the Dubois family lives within sight of one of France's oldest nuclear power stations. The cooling towers rise in the distance, as ordinary a feature of the horizon as steeples or silos. For Madame Dubois, the plant is neither abstract nor controversial — it is simply part of where they live. Her husband works there, as did her father before him, and the steady presence of the station has long been woven into the rhythms of local life.

On market days, work jackets with plant logos mingle with aprons and school uniforms; the plant's schedule sets a subtle rhythm for village life.

That rhythm did not emerge by accident. France's turn toward nuclear power began in the upheaval of the 1970s. After repeated oil shocks exposed the risks of relying on imported fuel, the French government launched the Messmer Plan — a sweeping effort to rapidly expand nuclear generation across the country. Over the next two decades, dozens of reactors were built with a clear objective: Secure energy independence.

The scale of that decision still shapes daily life today, with nuclear power providing roughly 70 percent of France's electricity — among the highest shares anywhere in the world.

Standardized designs and a national training pipeline created a workforce that could move from site to site, spreading know-how and consistency.

For families, this decision reshaped daily life. French electricity bills are stable and relatively low. Air quality is cleaner than in many neighbouring European countries. Nuclear jobs sustain local economies: schools, clinics, and even sports teams in towns like Flamanville or Gravelines are supported by the industry.

When heatwaves roll through, households notice that the grid holds; when neighbours across borders face shortages, France often exports power to help.

Public opinion has generally been favourable. Surveys show French families support subsidies for nuclear on par with renewables. Nuclear has become part of national identity — a point of pride, as ordinary as the TGV high-speed rail or the Eiffel Tower.

Families discuss reactors with the same matter-of-fact tone used for rail timetables: reliable, technical, unglamorous — and essential.

Challenges remain. The Flamanville 3 reactor, a next-generation European Pressurized Reactor (EPR), has been plagued by delays and cost overruns. Waste storage is an ongoing debate. Yet even with these issues, French leaders double down. In 2022, President Emmanuel Macron announced plans for at least six new reactors and investments in Small Modular Reactors (SMRs). The message to families was clear: Nuclear is not just history; it is the future.

Engineering schools report renewed student interest; apprenticeships in plant maintenance fill quickly; towns earmark land for future clean energy clusters.

For French families like the Dubois, nuclear is not feared or abstract. It is work, pride, and independence. It is the steady hum behind daily life — a reminder that national choices made decades ago still shape dinner-table conversations today.

When their son considers a career, the question isn't "Is nuclear safe?" but "Which role suits me best?"

Japan: A Legacy of Lessons

In a modest apartment in Sendai, the Tanaka family has adapted quietly to a changed reality. An emergency kit sits near the door — water, flashlights, spare batteries — checked and replenished as routinely as groceries. "Before," Mrs. Tanaka says, "we never thought about radiation at all." Now preparedness has become a habit, one more lesson absorbed into daily life.

Earthquake drills at school end with quiet reminders about evacuation routes. Families fold preparedness into routine the way others plan for snowstorms.

Japan's nuclear story cannot be told without March 11, 2011. A magnitude 9.0 earthquake struck. A tsunami followed. At the Fukushima Daiichi plant, backup systems failed. Explosions rocked reactors. Families fled in fear of radiation. For many Japanese, trust in government and industry was shattered that day.

Temporary housing became semi-permanent; family photo albums were left behind; communities scattered and re-formed elsewhere.

In the years that followed, Japan shut down all fifty-four of its reactors. Some were permanently decommissioned. Others sat idle, awaiting new safety approvals. To fill the gap, Japan imported fossil fuels — oil, coal, and liquefied natural gas. Bills rose sharply. Emissions spiked. Families paid more at the pump, in grocery stores, and on electricity bills.

Factories adjusted shifts to conserve power; convenience stores dimmed lights; summer "cool-biz" office attire became a necessity, not a novelty.

Gradually, under stricter regulations, Japan began restarting reactors. By 2025, about a dozen were back online, with more in the pipeline. But public opinion remains deeply divided. In some towns, families protested restarts, unwilling to risk another accident. In others, communities dependent on nuclear jobs welcomed the decision.

New safety requirements — higher seawalls, filtered vents, extended backup power — became visible symbols of promises made to the public.

The debate is deeply personal. For older generations, Fukushima is a trauma — a reminder of displacement and uncertainty. For younger families, it is a dilemma: how to balance safety fears with the reality of climate change and high costs. Surveys suggest perceptions are shifting, as more Japanese families see nuclear as a necessary tool against rising bills and carbon emissions.

Kitchen conversations often end with a careful sentence: "We must do both — be safer than before, and cleaner than before."

Japan's story is one of ambivalence. Families carry the memory of loss, but also the weight of future risks. Nuclear here is not just technology. It is memory, fear, and reluctant necessity.

Trust is being rebuilt the slow way: meeting by meeting, inspection by inspection, season by season.

Canada: Quiet Confidence

In the lakeside town of Kincardine, Ontario, the Noah family hosts a summer barbecue. Over the trees, the cooling towers of the Bruce Nuclear Generating Station loom large.

"It's part of our community," says Mr. Noah. "It's why we have good schools, good roads, and clean air."

Neighbours swap stories about apprenticeship programs and scholarships funded by plant contractors; the plant's outage schedule determines busy seasons for local diners and motels.

Canada's nuclear journey began in the 1940s with research and innovation, leading to the development of the CANDU reactor. Today, nuclear provides about 15 percent of Canada's electricity — and more than half of Ontario's.

Because CANDU reactors can use natural uranium and be refuelled online, Canadians often describe them as "workhorses" — steady and practical.

Ontario's reliance on nuclear enabled the province to phase out coal entirely by 2014. Families noticed the difference immediately. Smog days disappeared. Asthma rates in children dropped. Clean air became a quiet benefit of nuclear policy.

On summer evenings, parents point out the skyline: "Remember the haze? It hasn't been like that in years."

Jobs are another dimension. Refurbishing reactors at Bruce and Darlington created thousands of high-paying trades jobs. Families benefited from stable employment, strong local economies, and community investment. Nuclear jobs meant scholarships, sports sponsorships, and robust infrastructure.

Electricians, welders, pipe-fitters, operators — the energy transition here looks like skilled hands and long careers.

Public opinion in Canada is generally supportive, especially in regions that benefit directly. The federal government has embraced nuclear as part of its climate strategy. Investments in SMRs aim to bring clean power to remote northern communities, replacing expensive and polluting diesel.

Pilot projects pair SMRs with district heating and hydrogen production, promising warmth in winter and new industries year-round.

Canada's nuclear story also includes lessons in fairness. Indigenous communities have often borne disproportionate costs from energy projects, from uranium mining to hydro dams. Today, there is greater emphasis on consultation, consent, and equity. Families across Canada are learning that clean energy must also be just energy.

Partnership models — revenue sharing, training programs, and environmental monitoring led by local guardians — are becoming part of the blueprint.

For Canadian families, nuclear is steady, ordinary, and practical. It is not a subject of drama, but of quiet confidence — a technology that does its job, clearing skies and powering homes.

"Boring is beautiful," one parent jokes — meaning lights that stay on, air that stays clear.

South Africa: Blackouts and Ambitions

In Cape Town, the Mbeki family eats by candlelight during another round of "load-shedding" — scheduled blackouts to balance a fragile grid. Children study by lantern. Parents cook quickly before the power cuts again. Rolling outages are part of daily life.

Freezers thaw, traffic lights blink out, and small shops close early — a thousand tiny disruptions that add up to a heavy burden.

South Africa is the only African nation with commercial nuclear power. Its Koeberg plant, outside Cape Town, provides about 5 percent of national electricity and nearly all of the city's carbon-free power. For decades, leaders promised more reactors, but expansion plans stalled amid cost concerns and corruption scandals.

Meanwhile, aging coal stations falter; maintenance backlogs grow; diesel generators devour public funds meant for schools and clinics.

Meanwhile, families endure blackouts that stifle businesses, threaten health care, and sap morale. For many, nuclear represents hope: a technology that could deliver steady, reliable electricity in a country desperate for it. For others, nuclear is a symbol of mistrust — of elites making promises while households suffer.

Community meetings brim with practical questions: "How soon? How safe? How will this lower my bill?" The answers will shape whether nuclear feels like rescue or rhetoric.

The debate is not academic. For families like the Mbekis, reliable power means safe food, working hospitals, and jobs that depend on electricity. Without it, daily life is precarious. Nuclear, in this context, feels less like ideology and more like survival.

Parents tuck candles in drawers and power banks by the door, but what they really want is a socket that works every day.

South Africa's story reveals how nuclear is viewed differently in the Global South. Here, the question is not "Is nuclear too risky?" but "Can we afford to live without it?" Families who sit in darkness each night may see nuclear not as a luxury, but as a necessity.

Energy, here, is not an abstract climate lever; it is a lifeline to dignity.

United States: Polarized and Patchwork

In Illinois, the Ramirez family proudly notes that their state produces more nuclear power than any other in America. They see nuclear as clean, safe, and essential to fighting climate change. In California, the Brooks family joined campaigns to shut down Diablo Canyon, fearing accidents on a seismically active coast.

Between these households lies a continent of different grids, different politics, and different lived experiences with risk.

This is the U.S. story: patchwork and polarized. Nuclear provides about 20 percent of national electricity, but attitudes vary widely. In some regions, it is embraced as a climate tool. In others, it is resisted as dangerous.

State policies pull in opposite directions: zero-emission credits to save existing plants in some places, early closures in others; new SMR proposals in yet others.

The U.S. Navy has run nuclear-powered ships safely for decades, a fact families often cite in favor of civilian use. Yet memories of accidents like Three Mile Island linger. Pop culture — from superhero origin stories to post-apocalyptic movies — shapes perception as much as science.

Local trust matters: Where plants sponsor school programs and emergency drills are transparent, support tends to be higher.

The climate crisis is shifting perspectives. Young climate activists now argue for keeping existing reactors open. Families facing wildfires, hurricanes, and rising bills see value in steady, carbon-free power. States like New York and Illinois have passed policies to save reactors from closure.

At the same time, communities insist on stronger oversight, clearer waste plans, and tangible local benefits — "Clean and fair," as one town hall sign read.

But polarization runs deep. Some families view nuclear as progress, others as peril. The U.S. story reflects its culture: diverse, divided, and fiercely debated. Nuclear here is a mirror of America itself — powerful, contested, and unresolved.

The next chapter may be written in smaller steps: one SMR at a time, one community benefit agreement at a time.

Six Stories, One Conversation

Six countries. Six different nuclear stories. Germany's phase-out, France's embrace, Japan's trauma, Canada's quiet confidence, South

Africa's blackouts, America's polarization.

Add others and the pattern holds: Culture, memory, and geography bend the arc of energy decisions.

For families, these choices are not abstract. They are bills on the kitchen table, inhalers for asthmatic children, jobs in local plants, or candles lit during blackouts. Nuclear energy looks different everywhere, but it always intersects with daily life.

Behind every policy is a bedtime routine, a school drop-off, a paycheck — the ordinary moments energy makes possible.

Together, these perspectives remind us that energy is cultural. It is about values, trust, and fairness as much as kilowatts and costs. There is no single path. But in a climate emergency, families everywhere are asking the same question: How do we build a future that is safe, clean, and just?

The answers will be stitched locally — but the fabric, shared by all of us, must be strong enough to keep the lights on and the air clear for our children.

Family Action:

Explore how families in other countries power their lives — from France's steady reactors to South Africa's blackouts. Compare their choices to your own, and talk about what values guide each path. Understanding different stories builds empathy — and helps every family see their place in the global clean energy journey.

CHAPTER 14

Indigenous Rights and Energy Projects

Whose land, whose choice, whose future?

Energy and the Land We Share

Every source of energy begins with land. Dams stretch across rivers. Turbines rise on ridges. Mines carve into mountains. Even nuclear plants, with their dense fuel, require land for reactors, cooling towers, and carefully designed storage. For most families, land is seen in terms of ownership — property bought and sold. For Indigenous peoples, land is something deeper. It is memory, story, and identity. It is where ancestors are buried, where children learn traditions, where rivers are more than water — they are relatives.

That is why decisions about energy projects strike so deeply. When land is disrupted without permission, Indigenous families lose more than resources. They lose parts of themselves. For too long, energy has been developed on Indigenous territories without consent. Uranium mined without adequate safety protections. Dams built without consultation. Nuclear tests detonated in ancestral seas. These histories live in memory and shape the mistrust many Indigenous families feel today.

The shift to clean energy could be a chance to repair this history, but only if Indigenous rights are placed at the centre of the transition. The future cannot be just about cleaner technology. It must also be about justice.

Families in cities often encounter energy as a bill; Indigenous families meet it as a neighbour that can either respect or trample their home. A just transition, then, is not simply "new technology faster," but "new relationships first."

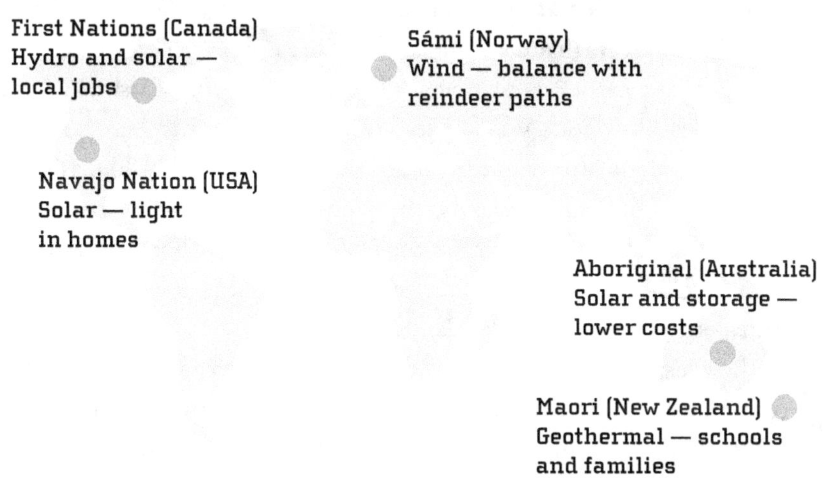

Fig. 15. Indigenous-led clean energy projects around the world

Lessons from the Past: Scars That Shape Trust

In Canada, Dene families remember the 1940s and '50s, when workers were hired to haul sacks of uranium ore out of the ground near Great Bear Lake. They carried it on their backs, unaware of the danger. Children played in the tailings, elders used the crushed stone for construction. Years later, cancers swept through communities. Families buried loved ones and wondered if the rocks had poisoned them.

In the American Southwest, Navajo families live with a similar legacy. Hundreds of uranium mines were abandoned after the boom ended. Wells ran contaminated, tailings blew across the desert, and livestock drank from poisoned springs. Grandparents tell stories of men who dug without masks or gloves, and of women who washed clothes heavy with radioactive dust. The grief of those years has never fully lifted.

Hydropower has often been presented as progress, but its history tells a more complicated story. In parts of Canada, Indigenous communities watched dams submerge traplines, hunting grounds, and burial sites that anchored generations. In Chile, similar projects displaced Mapuche families from rivers central to their identity. In many cases, assurances of consultation, compensation, or shared prosperity arrived late or not at all, leaving behind scars that continue to shape trust in new energy proposals.

The Pacific carries its own wounds. Families in the Marshall Islands recall the days of Cold War nuclear testing. They remember the blasts that turned night into day, the fallout that settled on skin like ash, the relocations to unfamiliar islands, the illnesses that followed. In French Polynesia too, families still grieve the consequences of bombs detonated over turquoise seas. For them, nuclear is not abstract science. It is a lived trauma, etched into bodies and landscapes.

These histories explain why Indigenous voices often approach new projects cautiously. Trust is fragile because it has been broken too many times.

Repair begins with acknowledgment. Community elders often say that before any proposal is discussed, the story of what happened must be told — aloud, with witnesses — so that new promises are made in the full light of old harms.

Diesel Dependency and Energy Poverty

Today, many Indigenous and remote communities live with another kind of burden: dependency on diesel. In northern Canada, fuel must be flown in or driven over temporary ice roads. Families pay some of the highest electricity prices in the country. Generators hum through the night, spewing soot into the air and fumes into homes. Parents ration fuel to stretch supplies through long winters. Children do homework by weak light when outages strike. Groceries spoil quickly in households where refrigeration is expensive and unreliable.

In the Pacific Islands, families face a different but related challenge. Diesel is shipped across oceans, driving up costs. Every rise in oil prices ripples through household budgets. And as seas rise, threatening entire islands, families know that their dependence on imported diesel makes them doubly vulnerable: to climate change and to economic shocks.

The irony is bitter. Many of these same communities sit closest to renewable or nuclear potential. Geothermal steam rises beneath their lands. Solar beats down across deserts. Uranium lies beneath their soil. Yet for decades, decisions about harnessing these resources were made far away, without their ownership or consent.

When communities do control the switch — a locally owned microgrid, a village-scale battery, a community heat pump network — the difference is visible: quieter nights, steadier fridges, dollars kept circulating on the land instead of leaving by tanker or truck.

The Principle of Consent

The United Nations Declaration on the Rights of Indigenous Peoples affirms the principle of free, prior, and informed consent. For families, this is not legal jargon. It is the same principle parents teach children: You don't make decisions that affect everyone without asking them first. You don't build on land that isn't yours without permission.

When projects ignore this principle, they repeat the mistakes of the past. When they honour it, they open the door to trust, fairness, and partnership. Consent is not an obstacle to clean energy. It is the foundation on which lasting solutions can be built.

Consent also means time: time to read documents in one's own language, time to consult among clans, time to bring in independent experts — not just company consultants — so that "informed" is real, not symbolic.

North America: Scars and New Models

In Ontario, seventeen First Nations came together to create the Wataynikaneyap Transmission Project. For decades, their communities relied on expensive diesel. Families woke at night to restart failing generators, children studied by candlelight, and outages during storms left elders vulnerable. Today, new transmission lines, owned in part by those very communities, connect them to the provincial grid. The change is more than technical. Parents see children doing homework under steady light. Refrigerators hum reliably, storing fresh food. Medical devices run without fear of failure. And the pride is palpable: This was not imposed; it was built together.

Similar ownership models are emerging around community wind and solar in the prairies and Arctic microgrids, where revenue-sharing agreements help fund language programs, youth training, and land stewardship. When families see direct benefits, skepticism turns into support and long-term care for their lands.

Contrast that with the Navajo Nation. In the dry hills of Arizona and New Mexico, families still live with the consequences of abandoned uranium mines. Elders recall the years when men carried ore home to build houses and fences. Only later did they realize the stone was toxic. Wells remain contaminated, livestock die, and illnesses linger. The U.S. government has pledged cleanup, but progress is slow. Families remain wary of any new nuclear projects. For them, reconciliation must begin with cleaning the old wounds.

In Alaska and Nunavut, diesel reliance drives up the price of milk and medicine. Yet village-led heat pump pilots and river-hydro projects show how consent-based design can lower bills while respecting caribou paths and fishing grounds.

Latin America: Rivers and Sunlight

Along the Bío-Bío River in southern Chile, Mapuche families still tell their children how the water once moved freely through the valley. Elders remember fishing grounds that fed entire communities, ceremonies guided by the river's seasons, and burial sites set carefully along its banks. When dams rose, those places disappeared beneath reservoirs. The river remained, but its meaning changed.

The Ralco Dam was built in the name of progress. Hydropower

promised clean electricity and national development, yet construction moved forward without the consent of the people whose lives were most entangled with the land. Entire valleys were flooded. Families were relocated. Compensation arrived late, if at all. What could not be replaced — cemeteries, sacred sites, ancestral memory — was rarely counted.

For many Mapuche families, the dam became more than an infrastructure project. It stood as a reminder that even energy labeled "clean" can carry heavy costs when decisions are made without listening.

Similar stories echo across the region. In Brazil, large hydro projects displaced Indigenous and riverine communities, submerging homes and fishing grounds that had sustained families for generations. Promises of jobs and development often faded once construction crews moved on, leaving altered landscapes and unresolved grievances behind.

These experiences shaped how Indigenous communities engage with energy proposals today. Clean power, many leaders insist, cannot be defined solely by carbon metrics. It must also respect land, consent, and continuity — or it risks repeating old patterns under new names.

Yet in Mexico's Sonora Desert, another story unfolds. Indigenous Yaqui communities have begun building solar farms on their land. Panels gleam under relentless sun, powering homes and schools. Revenues stay local. Youth are trained to install and maintain the systems. Elders speak of the sun as a gift they have always known, now harnessed with respect. For families here, solar power feels not like imposition but empowerment.

Across the Andes, Aymara and Quechua communities are testing wind and micro-hydro where grazing and water rights are protected by agreements written with elders at the table, not informed after the fact.

Oceania: Heritage, Resistance, and Survival

In New Zealand, Māori families gather around geothermal springs that have long been sacred. Today, these same springs generate clean power in Māori-led projects. Parents explain to children how the steam rising from the earth carries both ancestral stories and modern electricity. Innovation and tradition coexist. For Māori youth, these projects are lessons: Their heritage is not in conflict with progress but can guide it.

In Australia, Aboriginal families in Kakadu remember the Ranger uranium mine. Protests in the 1980s connected mining to cultural loss. Sacred sites were threatened, and land rights were ignored. Today, many Aboriginal communities approach energy projects with suspicion. But there are also stories of hope. In remote outback towns, Aboriginal-led solar farms now provide power to schools and clinics. Families take pride in energy that is both clean and under their control.

The Pacific Islands carry the heaviest scars. In the Marshall Islands, families recall Bikini Atoll, where nuclear tests forced relocation. They remember the ash-like fallout on skin, the illnesses that spread, and the generations who suffered. Nuclear, for them, means displacement and disease. Yet today, Marshallese families are pioneers in community solar, covering rooftops with panels to cut diesel dependence. For them, renewable power is not just a climate solution but a declaration: We will survive.

On atolls where land is a thin ribbon between ocean and lagoon, the phrase "energy security" means lights that stay on during storms — but also projects that won't wash away culture.

Africa: Between Promise and Risk

In Kenya's Great Rift Valley, Maasai families tend cattle on ancestral lands. The steam that rises from the ground powers geothermal plants. Some Maasai families welcome the projects, noting schools built, clinics funded, and jobs created. Others are more skeptical, pointing to grazing lands lost and water sources disrupted. Parents weigh the trade-offs: steady electricity versus traditions of herding and land stewardship. The debate is complicated, but at least it now includes Maasai voices — something absent in past projects.

In South Africa, families know nuclear through Koeberg, the continent's only commercial nuclear plant. In Cape Town, parents rely on it to keep lights on during endless load-shedding — rolling blackouts that disrupt school, work, and home. Outside the city, many rural families remain disconnected or underserved. When new nuclear plants are proposed, families ask hard questions: Will the benefits reach us, or will they remain concentrated in cities and elite hands? Trust is fragile, but the desire for reliable power is urgent.

Across the Sahel, village-scale solar microgrids run by women's co-ops are lowering kerosene use and creating evening study time for children — reminders that energy justice is also gender justice.

Scandinavia: Reindeer and Resistance

Across northern Scandinavia, Sami families continue a way of life shaped by reindeer migration routes that have guided movement across the tundra for generations. When wind farms were built along these paths, herds became unsettled and traditional migrations were disrupted. Sami communities organized protests, arguing that renewable energy could not override their right to cultural survival. Courts ultimately agreed, ruling that the projects violated Sami rights — a rare but significant acknowledgment that clean energy must also respect Indigenous sovereignty.

For Sami parents, the lesson for their children is clear: Climate action is necessary, but it must not repeat old patterns of erasure. Clean energy cannot come through "green colonialism." It must come with respect.

Developers who return to the table now bring biologists, herders, and translators — evidence that consent is a process, not a checkbox.

Toward Nuclear Partnerships?

For many Indigenous communities, nuclear remains tied to pain — uranium mining scars, waste debates, nuclear testing memories. Yet in some places, especially where diesel dependency makes life precarious, interest in nuclear is slowly emerging. Small Modular Reactors are being discussed as possible solutions for remote areas.

But any future nuclear projects must be built differently. They cannot be imposed from above. They must be co-designed, co-owned, and rooted in consent. They must bring jobs, training, and pride, not just electrons. Imagine Indigenous youth not as passive recipients of energy decisions, but as engineers, operators, and decision-makers of reactors in their own territories. This vision is distant, but possible — if trust is rebuilt with respect.

That also means clear waste plans from day one, community monitoring rights, benefit-sharing agreements tied to measurable outcomes, and the power to say no — all of which turn "projects" into partnerships.

Family Voices, Global Lessons

From Navajo deserts to Mapuche rivers, from Māori springs to the Sami tundra, Indigenous families remind us that energy is never only about kilowatts. It is about justice. Parents teach that land is not just property but relationship. Elders remind us that history matters, and that wounds take time to heal. Children remind us that every decision we make today will shape the inheritance they receive tomorrow.

The stories converge on a single truth: The clean energy transition cannot succeed without fairness. Families everywhere should hear this. Justice is not an obstacle to climate action. It is the very heart of it.

When projects are done with consent, lights turn on and trust turns up; when they are not, the price is paid in lawsuits, delays, and broken ties that dim communities for generations.

Justice as Energy

The climate emergency demands speed. But speed without justice repeats old harms. Families in Indigenous communities know this better than anyone. They ask the deepest question of all: whose land, whose choice, whose future?

The answer matters to every family, not just Indigenous ones. Because a world powered cleanly but unfairly is not the world we want to pass to our children. Justice is not optional. Justice is energy.

If the energy transition is a bridge to tomorrow, justice is the foundation. Build it first — and the bridge will hold.

Family Action:

Support Indigenous-led energy by learning protocols for engagement, respecting decision-making processes, and amplifying Indigenous voices in local planning.

CHAPTER 15

The Ethics of Nuclear Energy

Fairness, stewardship, and intergenerational voices

Balancing Risks, Rewards, and Responsibilities

Is it ethical to build nuclear plants if future generations must manage the waste?

This chapter is not about reactors or kilowatts — it's about values. Families everywhere are asking: Is it right to use nuclear energy? Do we owe it to future generations to avoid long-lived waste, or do we owe it to them to avoid a hotter planet? Who bears the risks, and who enjoys the benefits?

At the kitchen table, these questions sound less like policy and more like parenting: What promises do we make, and which costs do we refuse to pass along? Ethics turns bills and power plants into conversations about care, duty, and the kind of world our children will inherit.

Why Ethics Matters in Energy

When families talk about energy, the first questions are usually practical: How much does it cost? Will it keep us warm in winter? Is it safe? But beneath those questions lies something deeper. Energy choices are moral choices. They are about fairness, trust, and responsibility.

Ethics is the thread that ties together the things families feel most strongly about: the air in a child's lungs, the reliability of the outlet beside a medical device, the dignity of communities whose lands host our infrastructure. Even when we argue about numbers, we're really arguing about values.

A parent in Toronto may ask: "Do we have the right to produce waste that our grandchildren must guard?" A teenager in Berlin may counter: "Do we have the right to burn coal today and leave a hotter, more dangerous world for the future?" These are ethical dilemmas, not technical puzzles.

They ask us to compare two kinds of burdens: a small, contained responsibility versus a diffuse, planetary risk. Families weigh these every day in smaller ways — like choosing to store household chemicals safely rather than pouring them down the drain and letting the problem drift downstream.

Nuclear energy brings these dilemmas into sharp focus. It is low-carbon, reliable, and powerful. But it also creates long-lived waste and stirs memories of accidents and weapons. Deciding whether to embrace it means weighing benefits against risks — not just for today's families, but for generations unborn.

Because the benefits and the burdens arrive on different schedules, ethics asks us to keep two clocks in our heads at once: the urgent clock of the climate and the patient clock of stewardship.

Fairness: Who Gains, Who Bears the Risk?

In France, families enjoy stable bills and clean skies thanks to nuclear power. Jobs in plants support schools, clinics, and local sports clubs. The benefits are clear and widely shared.

Fairness here looks like reciprocity: Communities that host reactors also receive careers, training, and long-term investments. Children grow up seeing the plant not only as a skyline feature but as a sponsor of their school robotics team.

But in the Navajo Nation, families live with contaminated wells and abandoned mines from decades of uranium extraction. They bear risks and harms without having shared fairly in the benefits.

This is ethics as geography: When clean air is enjoyed in one place while harm is concentrated in another, "green" feels less like virtue and more like displacement. Repair must start where the harm accumulated.

Fairness means asking: Who breathes cleaner air, and who carries the burden of waste? Who receives jobs, and who loses land? Families know fairness instinctively. It's the same principle children argue about when dividing chores or slices of cake: Are we sharing the work and the reward equally?

Practical answers include benefit-sharing, co-ownership, and siting rules that treat consent as a gate, not a speed bump — turning fairness from a slogan into a structure.

Stewardship: What Do We Owe the Future?

Every parent thinks about inheritance — not just money, but the kind of world we leave behind. Energy choices are part of that inheritance.

We plant trees whose shade we'll never sit under; we label boxes so others can find what they need later. Stewardship is the quiet work of being a good ancestor.

Nuclear creates waste that lasts centuries. That raises a profound question: Is it ethical to leave future generations the task of guarding it? Some argue no — we must avoid any legacy that binds our descendants to danger. Others argue yes — because the alternative is leaving them an overheated planet with collapsing ecosystems.

Stewardship, then, isn't the absence of burden. It's the choice of which burden is smallest and most controllable. A sealed, monitored responsibility in one place may be more honest — and kinder — than an unsealed, invisible burden spread through the sky.

This is the central ethical paradox of nuclear power. The waste is real, but so is the climate crisis. Families must decide what kind of burden is more acceptable: a small volume of contained waste, or a vast atmosphere destabilized by fossil fuels.

Ethics invites design: robust repositories, community oversight, and intergenerational funds that pay for monitoring far into the future — so the promise to our grandchildren comes with resources, not just words.

The analogy is familiar to families who save for retirement or plant trees. They may never enjoy the fruits themselves, but they do it for those who come after. Stewardship is about responsibility to the unseen.

It is also about humility: writing instructions others can revise, leaving space for better technologies to emerge, and refusing to lock the future into our present blind spots.

Trust: Who Do We Believe?

Trust is fragile. In Fukushima, families were told the risk was under control — until it wasn't. In Chernobyl, information was hidden, and families paid the price. In Indigenous communities, promises of safety and compensation often went unfulfilled.

These memories live in family photo albums and oral histories. They do not vanish because a brochure says "state-of-the-art."

Ethics demands transparency. Families want honesty: about risks, about costs, about mistakes. Without trust, no technology — no matter how safe — will be accepted. Parents explain to children that trust is like glass: Once broken, it is hard to repair. The same is true for nuclear energy.

Trust is built with habits: open data, independent monitoring, community siren tests that feel routine rather than ominous, and leaders who say "We were wrong" as readily as "We were right."

Case Studies Around the World

France: Ethics of Pride and Progress

French families largely accept nuclear as ethical because it delivers stability, independence, and clean air. Nuclear was chosen democratically after the oil crisis, and families see it as a national project. Yet even in France, questions persist about fairness: Should rural communities host most of the plants while urban centres enjoy the benefits? Should waste be stored near one community for the sake of others?

The ethical work here is distribution: ensuring training, investment, and long-term safety oversight are shared as deliberately as electrons are.

Germany: Ethics of Refusal

German families saw nuclear as ethically unacceptable after Fukushima. Shutting down reactors was seen as a moral stand: better to accept coal and gas in the short term than risk another disaster. Yet as emissions rose and bills spiked, younger Germans began asking whether the moral high ground came at the cost of climate justice. For them, the ethics are shifting.

Their question is pointed: Is refusing one kind of risk forcing another onto children — smoke now, storms later?

Japan: Ethics of Memory

Japanese families live with deep trauma from Fukushima. For many, restarting reactors feels like betrayal. Yet rising costs and climate threats complicate the question. Is it ethical to reject nuclear if it means higher emissions and more disasters from climate change? Families are split, weighing safety memories against future risks.

Ethics here honors grief while insisting on clarity: If reactors return, they must return with stronger protections, community veto power, and honest contingency plans.

Canada: Ethics of Stewardship

In Ontario, families see nuclear as ethical because it eliminated coal, clearing smog and protecting health. But in Indigenous territories scarred by mining, nuclear feels unjust. Canada's ethical debate centres on stewardship: Can new projects be done fairly, with Indigenous voices leading and waste stored responsibly?

The emerging answer blends climate duty with consent: projects that are co-designed, co-owned, and audited by those most affected.

South Africa: Ethics of Equity

Families living with blackouts ask if it is ethical to reject nuclear when millions sit in the dark. For them, reliability is survival. Yet questions of corruption and fairness linger: Will nuclear benefits reach poor families, or enrich elites? Ethics here is about equity in who gains from power.

An ethical "yes" would mean lights on in townships, not just towers; contracts that train local youth; and safeguards that keep public money public.

United States: Ethics of Division

In America, nuclear splits families. Some see it as ethical duty in the climate fight. Others see it as reckless, tied to weapons and risk. The ethical divide reflects the country's polarization, where values of independence, safety, climate responsibility, and justice all collide.

Bridging that divide may depend less on arguments and more on practice: keeping existing plants safe and transparent, proving new designs in partnership with host communities, and pairing nuclear growth with visible pollution cuts where kids breathe.

Intergenerational Justice

Children today already carry climate anxiety. They ask parents why summers are hotter, why fires rage, and why storms are stronger. They want to know if adults are acting fast enough.

They don't ask for perfect; they ask for courage and care, for choices that make tomorrow feel inhabitable.

Nuclear raises a hard but necessary question: What do we owe them? Is it fair to let fear of accidents or waste outweigh the urgency of climate action? Is it fair to gamble their safety on fossil fuels because we are unwilling to make hard choices today?

Intergenerational justice reminds us that delay is itself a decision — one that allocates risks to the young and to the not-yet-born.

Parents often put it simply: "We wear helmets not because we expect to crash, but to stay safe if we do. Nuclear energy works the same way for the climate. It isn't perfect, but it helps protect our future."

And like a helmet, it works best as part of a set: seat belts (efficiency), better roads (grids), and rules of the road (good policy).

An Ethical Balancing Act

The ethics of nuclear energy are not black and white. They are shades of responsibility, fairness, trust, and care.

They live in specifics: who decides, who benefits, who watches the watchers, and how we say "Stop" when promises are not kept.

Families must weigh real risks against real benefits. They must decide what burdens are acceptable to pass on, and which are too heavy. They must balance the needs of today with the rights of tomorrow.

The ethical centre of gravity shifts toward solutions that are transparent, locally partnered, and reversible where possible — so that future generations inherit options, not just obligations.

In the end, ethics is not about perfection but about priorities. Do we value cleaner air for children today? Do we value a stable climate for grandchildren tomorrow? Do we believe communities deserve a voice in choices that shape their lives?

If our answers are yes, then the path forward becomes clearer: Cut the smoke now, contain the waste well, share the gains fairly, and keep faith with the people who live where power is made.

Nuclear energy forces these questions more starkly than most technologies. And in facing them, families are reminded that energy is never just technical. It is moral. It is about what kind of world we want to leave behind.

The light in a child's bedroom is powered not only by steam and wires, but by our promises — to keep that light clean, steady, and just for those who come after.

Family Action:

Weigh ethics at home by listing who benefits and who bears risk across options, then decide what safeguards matter most for your family.

CHAPTER 16

Energy Justice and Who Benefits

Who gains, who pays, who decides?

Balancing Benefits and Burdens

Energy justice looks closely at how power systems affect different groups of people, asking questions that echo through homes and communities everywhere. If ethics asks, "Is this right?" justice asks, "Right for whom?" It is about distribution: who gains the benefits of energy systems, and who bears the costs when things go wrong.

Families can relate to this instinctively. When dinnertime arguments erupt over who gets the last piece of pie or which sibling clears the table, the principle is the same. Fairness is not an abstract principle; it is daily life. Energy systems, vast as they are, run on those same questions of fairness.

Why Justice Matters in Energy

When families think about justice, it often begins with the smallest of daily moments. Children argue over whose turn it is to take out the garbage, or which sibling gets the coveted window seat on a car trip. Parents debate whether chores, allowances, and privileges are shared fairly. At its core, justice is about how we divide burdens and benefits — who works, who waits, who enjoys the prize.

Energy systems, although vast and technical, operate on exactly the same principle. Power plants, transmission lines, and government policies touch every home, yet they do not touch all homes equally. Some families breathe cleaner air because they live upwind of a coal plant, while others cough through sleepless nights in the shadow of smokestacks. Some neighbourhoods pay higher rates because their housing is old and leaky, while affluent districts with efficient insulation enjoy lower bills. Still others — often the poorest — live near landfills for coal ash or beside pipelines that carry risks without delivering many benefits.

Justice matters in energy because energy itself is the invisible skeleton of modern life. It heats the water for a baby's bath, powers the incubators in hospitals, and charges the phones that connect loved ones in a storm. When access or quality is unfair, the imbalance ripples into education, health, and opportunity. A child who cannot study at night because of frequent blackouts falls behind in school. An elderly neighbour whose electricity is cut during a heatwave faces mortal danger. These are not abstractions; they are real-life kitchen-table crises.

True energy justice therefore demands more than clean technology. It requires fair processes and representation. The communities most exposed to pollution or most vulnerable to outages must have real seats at the table where decisions are made. Justice is not a luxury to consider after engineering problems are solved; it is a guide for every choice from where to build a wind farm to how to set household rates.

The Unequal Burden of Fossil Fuels

Families living near coal plants know what injustice smells like. In West Virginia, mothers wipe black dust from windowsills each morning. Children carry inhalers to school and measure their days by asthma flare-ups. In India's coal belt, parents describe skies so hazy they cannot see the horizon; health clinics fill with respiratory patients.

The same story repeats across continents. In Poland's Silesia region, miners' wages support towns even as children wheeze from smog. In Nigeria's Niger Delta, oil spills coat mangroves and drinking water while profits flow to distant capitals. Globally, the World Health Organization estimates that millions die prematurely each year from air pollution, much of it tied to fossil fuels.

Yet the families who suffer the asthma attacks or watch crops fail from acid rain are rarely the ones profiting from the electricity or the gasoline. Justice means recognizing that not all lungs are equal in the fossil-fuel era. Some families live downwind of smokestacks while others live far away, enjoying the benefits without the costs.

The Energy Burden on Low-Income Families

In the United States, energy justice shows up every month when the bills arrive. Wealthier households typically spend about 3 percent of their income on electricity and heat. Low-income families often spend 10 percent or more. Parents must choose between paying the light bill and buying groceries. Children study under lamps that cost more per hour to run than their classmates' lights.

This "energy burden" surfaces elsewhere too. During the 2022 gas crisis, families across Germany and the United Kingdom saw bills double or triple. Wealthier households could absorb the costs or invest in efficiency upgrades — new windows, better insulation, smart thermostats. Poorer households wore coats indoors, skipped meals, or fell into debt. Fuel poverty became not just an economic term but a lived reality.

In South Korea, families in older apartments pay proportionally more because insulation is poor and meters are outdated. In Arctic regions of Alaska and northern Canada, Indigenous families pay some of the highest electricity prices on the continent because diesel fuel must be transported across vast distances, sometimes by air or hauled over ice roads. Justice is not only about technology; it is about affordability and dignity. Families deserve light and warmth without sacrificing health or food.

Rural and Urban Divides

Energy injustice often follows the fault lines between countryside and city, a gap that stretches across continents. In many nations, city lights blaze late into the night while rural families still plan their lives around the uncertainty of power cuts.

Consider South Africa, where rolling blackouts — known locally as "load-shedding" — darken homes on a daily schedule. Urban households, though inconvenienced, often cushion the blow with rooftop solar panels, private generators, or battery backups. Restaurants keep serving meals, and children finish homework under steady LED lamps. But in small farming villages and isolated townships, families endure hours of darkness with no fallback. Food spoils, water pumps fail, and schoolwork stops when candles burn out. The same national blackout feels profoundly different depending on your postal code.

A similar story plays out in rural India, where families may wait half the day for the single hour when the grid sends power to their village. Parents rise before dawn to pump water when electricity is finally available, while city hospitals a few hundred kilometres away never lose power at all. In the United States, many remote Alaskan communities must fly in diesel by plane or haul it across winter ice roads. Every gallon of fuel raises the price of heating to levels unimaginable in urban centres, where natural-gas pipelines and nuclear plants keep prices steady.

Even in wealthy countries, the gap between rural and urban energy costs remains wide. In northern Canada, many Indigenous households pay three times more per kilowatt-hour than people in cities because transmission lines don't reach their communities, forcing them to rely on expensive diesel power. Meanwhile, suburban families farther south use efficient heat pumps powered by low-cost hydroelectricity.

This disparity is not just about geography. It is about fairness. Families who live farthest from big hospitals and supermarkets — the ones who need reliable energy the most — are often the last to receive it. Energy justice means ensuring that a parent in a mountain village can keep a newborn warm just as surely as a parent in a high-rise apartment downtown.

Renewables and Unequal Benefits

Clean energy is often framed as inherently fair, yet experience shows that outcomes depend on how projects are planned and shared. In Brazil, large dams displaced Indigenous and riverine communities. In Chile, hydro projects reshaped sacred landscapes. In northern Scandinavia, wind farms intersected with Sami reindeer migration routes. These cases reveal a common pattern: When communities are excluded from decisions, even low-carbon energy can reproduce old inequalities under a new name.

At the same time, families in wealthy suburbs of California or Germany proudly display solar rooftops, enjoying lower bills and generous tax credits. Renters, apartment dwellers, and low-income households are left behind, excluded from benefits they help subsidize through taxes.

Clean energy is not automatically just; it must be designed to include everyone. Community solar gardens, shared ownership programs, and targeted subsidies can help ensure that the renewable transition doesn't widen old gaps but instead closes them.

Nuclear Justice Questions

Nuclear energy brings its own unique set of justice challenges — questions that linger long after the lights turn on. Families in Ontario, Canada, breathe cleaner air today because nuclear plants eliminated coal, drastically reducing asthma and smog. French households enjoy stable bills and one of the cleanest electrical grids in the industrialized world, thanks to decades of nuclear commitment. These are undeniable benefits.

But not every family shares them equally. In the American Southwest, Navajo communities still live with the scars of uranium mining: contaminated wells, poisoned sheep pastures, and a legacy of illness. Many of these mines operated without proper consultation or consent, and their cleanup remains incomplete decades later. In other parts of the world, rural towns host long-term waste repositories while urban centres consume most of the electricity. A village near a proposed repository in France or Finland may feel that it shoulders risks for the sake of distant cities that reap the rewards.

Justice in the nuclear realm therefore extends beyond engineering safety. It asks who decides where plants or waste sites are built, who monitors them, and who benefits from the jobs and revenue they generate. Are local residents offered real partnership — ownership stakes, revenue sharing, training programs, or merely promises of compensation? Are Indigenous voices included from the very start, with legal power to say no?

The new generation of Small Modular Reactors (SMRs) now being proposed for remote regions tests these questions all over again. Support will depend on whether communities can truly co-design projects, not simply host them. A just nuclear future is possible, but only if the people who live closest to the reactors also hold the strongest voice in shaping them.

Families and Climate Justice

The climate crisis deepens every inequity. Families in Bangladesh, who contributed least to global emissions, face rising seas that swallow farmland. Families in the Pacific Islands watch ancestral graves sink beneath waves strengthened by warming oceans. Meanwhile, wealthier countries debate how quickly to cut carbon while enjoying the luxuries that fossil fuels built.

For children, climate justice is intergenerational. They ask why they must inherit the consequences of decisions they did not make. Parents must answer whether fairness stretches not only across borders but across time.

Justice also means adaptation funding: helping vulnerable families build seawalls, move to safer ground, or install cooling systems as heatwaves grow fiercer. Without that support, the poorest pay twice — first through the impacts, then through the cost of rebuilding.

Analogies of Fairness

Families do not need graduate courses in economics to grasp the essence of energy justice; they live its metaphors every day. A fair dinner table means every child receives a portion, not just the quickest to reach for the serving spoon. A fair inheritance means siblings share both the treasures and the chores that come with keeping a family home. A fair classroom means rules apply to everyone, not just to the quiet students who never protest.

Energy is no different. A power system that gives some neighbourhoods pristine air while others choke on smoke is like a meal where one child feasts and another goes hungry. A climate policy that delays action and leaves the costs to future generations is like a parent who eats the harvest but leaves the field barren for their children.

Parents teach toddlers to take turns with toys; grandparents tell stories about helping neighbours during storms. These small lessons form a moral foundation. Energy justice is the grown-up version of that same teaching, scaled up to nations and generations. When families see energy through this lens, they understand that fairness is not a lofty slogan but a daily practice — one that demands participation, empathy, and the courage to share both the work and the reward.

Justice as Participation

Energy justice is not about perfection; it is about participation and balance. Families must have voices in decisions that affect their land, their air, their bills, and their futures.

That means town hall meetings in local languages, community ownership stakes in wind or solar farms, and clear legal rights to accept or reject projects. It means data that is public and understandable, so families can see whether promises are kept.

Who benefits and who bears the cost is not a technical question. It is a moral one. Families everywhere deserve to share in the light of clean power without being left in the shadow of its costs.

A Shared Table of Power

In the end, justice in energy is the same as justice in family life: Everyone should have a place at the table, a voice in the conversation, and a fair share of the benefits. Only when that happens will the power that warms our homes also strengthen the bonds of fairness that hold societies together.

Expanding renewable grids, safer nuclear plants, and smarter policies can all move us closer to that vision — but only if the voices of those most affected are heard. Families understand this truth instinctively. A just energy future is not a technical blueprint; it is a moral commitment to each other, built one household, one neighbourhood, and one generation at a time.

Family Action:
Look for community energy opportunities through co-ops or municipal programs and ask how revenue is reinvested locally.

CHAPTER 17

Public Engagement and Trust

Why people matter as much as reactors

Openness matters as much as safety

Technology alone doesn't build trust. *People do.*

Every family already understands this truth. A house can be fitted with the finest locks and alarms, but it is the daily honesty among its members that keeps it a home. In the same way, the cleanest power plant, the most carefully engineered nuclear reactor, or the most dazzling solar farm cannot succeed without the confidence of the people living beside it.

Think of how families navigate ordinary life. Parents tell children where they are going and when they will return. Neighbours let one another know if a tree might fall across a shared fence. A marriage thrives not on perfection but on conversation, on the small rituals of keeping one another informed. Energy projects — especially ones as complex and long-lasting as nuclear — depend on this same rhythm of openness.

Communities that host power plants are not merely passive backdrops. They are homes where children play in backyards, where grandparents watch over gardens, where parents worry about air quality, property values, and the next generation's opportunities. When those communities feel heard, the wires and turbines around them become part of everyday life. When they feel ignored or deceived, even the most advanced technology feels like a threat. The lesson is as old as family life itself: Trust is the invisible infrastructure on which everything else rests.

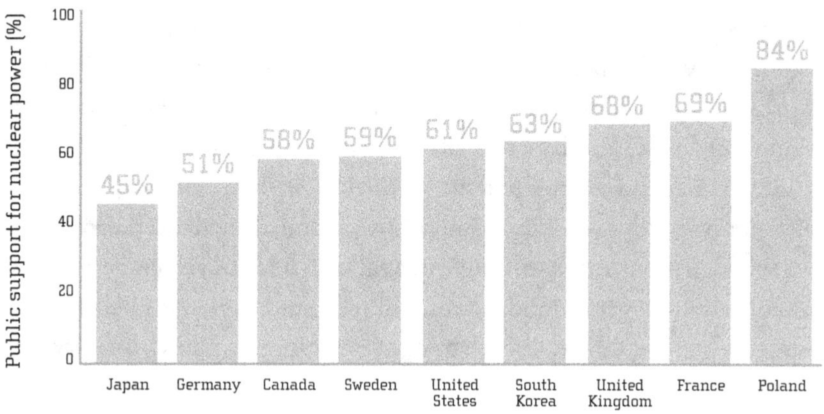

Fig. 16. Public support for nuclear power for selected countries

The Fragility of Trust

Every household knows that trust is easy to lose and painfully slow to rebuild. A single broken promise, a secret kept too long, a lie told "for our own good" — all can undo years of reliability. Children quickly remind parents when promises aren't kept; parents teach children that honesty is the foundation of lasting relationships.

Energy is no different. Communities trust power companies to keep the lights on and to keep them safe. Nuclear energy, in particular, relies not only on redundant safety systems of steel and concrete but on a social safety net of confidence. A family living near a reactor must believe that, if something goes wrong, they will be told the truth quickly, clearly, and completely.

Researchers call this confidence the "social license to operate." Without it, even flawless engineering can fail in the court of public opinion. When trust is strong, nuclear plants feel like steady neighbours — reliable, almost invisible. When trust is broken, even the safest reactor cannot reassure families. After major energy accidents, studies consistently show that the loss of confidence often drives more evacuations, psychological stress, and economic damage than the physical event itself.

When Trust Breaks: Lessons from Accidents

History shows how quickly that invisible infrastructure can fracture.

In 1986, the people of Pripyat, a lively Soviet city built to house workers of the Chernobyl plant, woke to a blue glow on the horizon and rumours of a fire. For days, officials insisted nothing serious had happened. Children went to school, parents worked their shifts, and families spent afternoons outdoors while a radioactive cloud drifted overhead. Later, when evacuations came without explanation, the betrayal felt deeper than the physical danger. Grandparents still tell those stories at kitchen tables across Ukraine and Belarus, shaping how young people view every new energy proposal.

A quarter century later, Japan endured the triple disaster of earthquake, tsunami, and the Fukushima Daiichi meltdown. Here too the greatest wound for many families was not radiation exposure — which later studies showed was lower than feared — but the confusion and silence in the crucial first days. Parents recall waiting in gymnasiums for buses that never arrived, wondering whether tap water was safe for their babies' bottles. Some still keep emergency bags by the door, a quiet testament to the trauma of not knowing.

Even far from reactors, these events ripple outward. German families watched the Fukushima footage and pressed their government to accelerate its nuclear phase-out. In the United States, neighbourhood groups near aging plants demanded new safety inspections. Psychologists who studied both Chernobyl and Fukushima concluded that the deepest scars were psychological: stress, dislocation, and a lingering belief that authorities value secrecy over candour. A single breach of honesty, they found, can echo through generations like an aftershock.

When Trust Holds: Stories of Openness

There are also places where openness has strengthened the bond between technology and community.

In Ontario, Canada, nuclear operators hold regular open houses in towns near their plants. Families tour control rooms, meet engineers, and handle training equipment. Children see how a reactor works and hear about safety drills. Parents hear directly about emergency plans and independent monitoring. Transparency builds confidence.

Finland provides another model. The Onkalo repository for nuclear waste — one of the first of its kind — is being built deep underground after *years* of public dialogue. Nothing was hidden: community meetings, local advisory councils, and citizen site visits became routine. As a result, Finland is now one of the few places in the world where a permanent waste facility enjoys broad public support. Families there believe that their voices shaped every step.

Sweden tells a similar story. Over three decades and more than five thousand public meetings, residents debated every detail of their proposed repository. Local families eventually voted in favour because they felt heard at every turn. In the United States, Illinois has kept aging reactors running in part because plant operators invite schools, neighbourhood groups, and journalists inside for regular tours and question-and-answer nights. When residents know the people behind the technology, abstract fears shrink.

Global Perspectives on Engagement

Other nations reveal how fragile, and how powerful, engagement can be.

In South Africa, proposed nuclear expansions have met deep skepticism. Families still remember corruption scandals in the energy sector and daily blackouts. Promises of reliability ring hollow when the grid fails every evening. Without transparency and genuine dialogue, even a well-designed project struggles to gain acceptance.

In France, nuclear power is woven into national pride. Families see it as part of independence and technological identity. Yet even here, engagement matters. When new projects face delays or unexpected costs, open communication keeps skepticism at bay. Silence would erode confidence; dialogue preserves it.

Across Asia and the Pacific, engagement is inseparable from history. Indigenous communities in Australia and the Marshall Islands recall uranium mining and nuclear testing imposed without consent. Today, when governments propose new partnerships, elders ask: "Why should we trust you now?" Trust can only be rebuilt with action — genuine consent, shared ownership, and visible respect.

The same lesson appears in renewable projects. When offshore wind developers in Denmark held hundreds of neighbourhood meetings and offered partial ownership to residents, public acceptance soared. Where those conversations were skipped — as in some early U.K. projects — opposition hardened, even to clean power.

Families at the Table

True engagement is never a single meeting or a glossy brochure; it is a continuing conversation that treats families as full partners in shaping their own energy future. Parents everywhere want clear, practical answers to questions that touch daily life: Is our drinking water safe if there is an accident? How will we be alerted if something goes wrong? Will our children find good jobs in this project, or will the benefits flow elsewhere? Trust grows only when these questions are met with patient explanations, site visits, and a genuine chance to influence decisions. When officials hide behind jargon or make choices out of public view, suspicion hardens and even the best-engineered project feels imposed.

Communities around the world show how powerful real participation can be when it becomes a habit instead of a performance. In Alaska, villages considering Small Modular Reactors have created local energy councils where elders, teenagers, and engineers sit side by side, translating technical plans into native languages and cultural traditions so that every resident understands the stakes. Finland offers another example: Residents near the Onkalo waste repository have spent decades in dialogue with planners, requesting independent monitoring and receiving regular updates. Parents there now describe the facility not as a mysterious "other," but as something they helped design and can hold accountable. On the windy coasts of Denmark, fishermen and farmers are invited into early planning sessions for offshore wind farms, and their concerns about fishing routes and scenic views directly shape where turbines stand. Because these conversations begin long before construction and continue long after, acceptance is remarkably high, and conflicts are rare.

Experiences like these show that families do not merely need to be consulted; they need to be included from the first sketch of a blueprint to the final ribbon-cutting. Participation transforms an energy project from something done *to* a community into something done *with* it. It allows parents to look their children in the eye and say, "We helped shape this, we understand how it works, and we can hold it accountable." That sense of shared ownership is what turns technology from an external force into a neighbour — and it is the surest foundation for the lasting trust any energy project needs.

Analogies Families Understand

Because energy systems can seem abstract, families often reach for everyday comparisons to make sense of them.

Trust in energy is like trust in a doctor. You agree to a treatment — even one with side effects — because you believe the doctor will explain every risk and stay with you through recovery. If the doctor hides information, confidence collapses no matter how advanced the medicine.

It is like the teacher who admits when she does not know an answer but promises to find it. Honesty builds respect, even more than expertise. Or the neighbour who warns you that your fence is leaning before your children get hurt; secrecy breeds resentment, but openness builds community.

Think of a family budget: When parents share the numbers with their children, everyone understands the trade-offs, and no one is blindsided by surprise expenses. Energy planning can work the same way — open books, clear priorities, and a sense that every household's needs are part of the equation.

Even the small rituals of daily life carry lessons. A parent who tells a child when they will be late coming home, a teenager who texts if plans change, a grandparent who explains why a family heirloom is fragile — all these acts of transparency create the confidence that holds a family together. Energy projects must practice the same everyday honesty if they hope to earn and keep the public's trust.

The Human Side of Safety

Engagement and trust are not extras; they are as essential as control rods or cooling systems. Families will only accept nuclear energy — or any major energy project — if they believe they are respected partners, not passive recipients.

The lesson is simple but profound. Safety is not just steel and concrete. It is honesty and dialogue. Families can forgive mistakes if the truth is told promptly and completely. They cannot forgive secrecy.

From local town halls to international treaties, the energy transition will only succeed if the people who live with its consequences have real power in shaping it. Engineers say redundancy keeps technical systems safe; in society, public trust plays the same role. It's the backup that keeps everything steady when surprises come

In the end, nuclear energy's greatest safety feature is not technology but trust. Families must be at the table, not as afterthoughts but as equal voices. Only then can the promise of clean, reliable energy rest on a foundation as strong as any reactor vessel — the foundation of trust.

Family Action:

Build trust locally by attending open houses, reading safety reports, and inviting experts to community Q&As where residents can ask hard questions.

CHAPTER 18

Culture and the Moral Case for Clean Power

How stories shape our sense of right and wrong

Confidence Grows Through Engagement

Energy isn't only wires and turbines. *It's also the stories we tell.* When families talk about energy, the conversation usually begins with the practical: the size of the electricity bill, the reliability of the grid, the air their children breathe. But beneath these everyday concerns is culture — the stories we tell ourselves about who we are, what we value, and how we decide what is right or wrong.

Culture explains why French families see nuclear energy as a source of national pride, while German families see it as a moral risk. It explains why Indigenous elders speak of rivers and mountains as relatives to be respected, while engineers speak of them as resources to be harnessed. It explains why Pacific Islanders describe climate action as survival, not choice.

Culture shapes morality, and morality shapes energy choices. To understand why families around the world make such different decisions about nuclear power and clean energy, we have to look not just at kilowatts or emissions, but at the *stories that guide them*.

Anthropologists call these stories "energy cultures": shared habits and beliefs that tell families what is acceptable, desirable, or forbidden long before a policy vote or a utility bill arrives.

France: Nuclear as Identity

In France, nuclear energy is not just tolerated. It is celebrated, woven into national identity.

For the Dubois family, the view from their backyard carries meaning beyond geography. The cooling towers outside Chinon represent continuity — work passed from one generation to the next, and a shared sense of contribution to the country's energy needs. Madame Dubois speaks of the plant with quiet pride, not as a symbol of risk but of responsibility: a way their family helps keep homes lit, food fresh, and emissions low across France.

That legacy carries emotional weight. In the wake of the oil crises, the Messmer Plan was presented not only as an engineering solution but as a statement of principle. By committing to domestic nuclear power, France sought to ensure that households would no longer be vulnerable to distant conflicts or foreign suppliers. For many families, the choice came to symbolize responsibility and self-reliance — a promise that energy security was something the nation owed its citizens.

Over time, this decision became culture. Children went on school trips to EDF visitor centres, where they learned about reactors as symbols of progress. Families grew accustomed to low bills and clean air compared to neighbours in coal-heavy Germany or gas-dependent Italy. Pride grew not only from independence but from success: France cut carbon emissions while keeping power reliable.

Yet pride does not erase problems. Families debate the delays at the Flamanville EPR project and the fairness of storing waste in rural areas far from Paris. Environmentalists worry about centralization and overreliance. Still, the cultural story remains strong. For most French families, nuclear is not just energy. *It is identity — proof that France can make bold choices and carry them through.*

Germany: A Moral Rejection

Cross the border into Germany, and the cultural story changes completely.

For households like the Schneiders', energy choices are rarely discussed in technical terms. Their home is heated with gas and powered by a mix of renewables and coal, but the deeper reasoning runs elsewhere. When Lena presses her parents about nuclear energy, her father explains that for many Germans, the issue was never only about emissions. Nuclear came to represent a line the country chose not to cross — a decision shaped by history, memory, and a collective sense of responsibility.

Germany's nuclear debate is moulded by culture as much as engineering. The anti-nuclear movement began in the 1970s, tied to Cold War fears of weapons and accidents. Protests swept the country in the 1980s, uniting environmentalists, church groups, and ordinary families. The Green Party rose on the promise of a nuclear-free Germany.

Over time, opposition to nuclear energy in Germany hardened into something deeper than policy preference. What began as protest evolved into principle. Nuclear power came to symbolize a broader unease with technologies perceived as uncontrollable, irreversible, or imposed without consent. For many families, rejecting nuclear was not about weighing kilowatts or emissions, but about affirming a moral boundary — a collective decision to err on the side of caution, even when the trade-offs were uncomfortable.

Families embraced the *Energiewende*, the energy transition. Rooftop solar panels and wind turbines became cultural symbols, displayed proudly as acts of citizenship. For German families, producing clean energy from wind and sun felt not just efficient but ethical, aligned with cultural values of caution, responsibility, and environmental stewardship.

But reality complicated the picture. Coal returned to fill gaps when renewables faltered. Imported gas, especially from Russia, became essential. When war cut off supplies, bills skyrocketed. Families found themselves asking new questions. Was it truly moral to reject nuclear if it meant burning more coal? Was refusing reactors a matter of principle, or a choice that hurt both climate and family budgets?

Germany's story shows how culture can shape morality even when it collides with practical outcomes. For many families, rejecting nuclear still feels right, even if it makes life harder. For others, especially the young, the moral compass is shifting back toward pragmatism.

Japan: Trauma as Memory

For Mrs. Tanaka, the legacy of Fukushima is not measured only in policy or infrastructure, but in awareness that never quite fades. Radiation, once an abstract concept, lingers at the edge of thought, shaping how safety is imagined and discussed. It is not panic that remains, but memory: a quiet presence that influences how risks are judged long after the immediate danger has passed.

Japan's relationship with nuclear is haunted by memory. Hiroshima and Nagasaki scarred the national psyche, embedding radiation in the cultural imagination as terror and trauma. Fukushima in 2011 deepened that wound. Families fled homes with minutes' notice. Children carried only a toy or a pet. Elders refused to leave ancestral land. Evacuations stretched for months. Trust in government collapsed as information trickled out slowly.

For many families, nuclear is not an abstract debate. It is lived experience, tied to stories of betrayal and loss. Parents tell children about March 11, 2011, as a cautionary tale. "This is why we must be careful," they say, "because the world can change in an instant."

Yet Japan faces energy realities that make the debate painful. The country imports most of its fuel, paying high prices and emitting carbon. Families endure high bills and rising climate risks. Younger generations, anxious about climate change, quietly ask: Is it right to keep rejecting nuclear when the alternative is worse?

The moral case in Japan is torn between memory and necessity. Culture says "Never again." Climate says "We need every tool." Families live in the tension, uncertain which story will guide the future.

Indigenous Traditions: Energy as Relationship

In Indigenous cultures around the world, morality is framed not by kilowatts but by relationships.

In Canada, Cree and Dene elders tell children that rivers are alive, that fish and moose are kin. When hydro dams flood traplines, families see it not just as economic loss but as spiritual harm. In New Zealand, Māori families view geothermal springs as *wāhi tapu* — sacred places. Harnessing them for electricity requires rituals and guardianship, ensuring the land is respected as ancestor, not exploited as commodity.

In Sami households, children learn that energy decisions are inseparable from relationships — to land, animals, and ancestors. Wind turbines that interrupt reindeer migrations become lessons in how even well-intentioned projects can cause harm when they ignore lived knowledge. Similar stories echo far from Scandinavia. In the American Southwest, Navajo families speak of uranium mining that contaminated water and soil, leaving behind illness and loss. For many Indigenous communities, energy is not judged only by carbon counts, but by whether it honors the bonds that sustain life.

These traditions remind us that morality is not just about efficiency or safety. It is about reciprocity. If energy takes without giving, it is immoral. If energy honours relationship, it can be just. For Indigenous families, the moral case for clean power begins with *respect and shared stewardship.*

Pacific Island Cultures: Climate as Obligation

In the Marshall Islands, families tell stories of lost atolls. Elders teach children about land that may soon vanish beneath rising seas. In Kiribati and Tuvalu, parents worry about where their children will live in twenty years. For Pacific Islanders, climate change is not an abstract concern. It is a matter of survival.

This shapes morality in profound ways. Cutting carbon is not seen as an option but as a duty — an obligation of justice. Families frame climate action as moral debt owed by big polluters to small islands. Nuclear legacies compound this. In Bikini Atoll, families were forced from their homes during U.S. nuclear testing. Generations later, cancers and contamination remain. For these families, nuclear is remembered with bitterness, while renewable solar and wind are embraced as survival strategies.

Yet even here, some families quietly ask: If nuclear could cut global carbon faster, should it be reconsidered? The debate is not settled, but the moral compass is clear: Whatever choices are made must prioritize survival. For Pacific cultures, the moral case is urgent and absolute.

Religion, Ritual, and Energy

Across the world, religion shapes how families frame morality in energy.

Catholic families draw on encyclicals like "Laudato Si'," which call for stewardship of creation. Parents explain to children that wasting electricity is not just costly but sinful — a failure to honour God's gift. Muslim families speak of *khalifa*, humanity's role as guardians of the Earth. Turning off unnecessary lights becomes an act of respect, not just thrift. Hindu families revere the Ganges as sacred. Polluting rivers becomes not only a technical problem but a moral one.

Jewish traditions invoke *tikkun olam* — repairing the world — as a mandate for climate action. Buddhist teachings on interdependence frame clean energy as a path to reduce suffering for all beings.

These religious traditions reinforce the moral case for clean energy. They remind families that electricity is not just mechanical but spiritual — a gift to be used with care.

Stories at the Dinner Table

Culture is passed down in stories. Around dinner tables, parents explain energy choices in moral language.

In France, a father says, "Nuclear keeps us free." In Germany, a mother says, "Wind and solar keep us safe." In Japan, a grandmother says, "Fukushima taught us caution." In an Indigenous home in Canada, an elder says, "The river is our relative; we must not harm it." In the Marshall Islands, a parent says, "If the world keeps burning coal, we will lose our home."

Children absorb these lessons. They grow up seeing energy not just as wires and bills but as values and identity. The moral case for clean power is taught not in textbooks but in kitchens, living rooms, and bedtime stories.

The Moral Case for Clean Power

So what is the moral case? At its heart, it is simple. Families everywhere want to pass on a world where children can breathe clean air, drink safe water, and live without fear of climate collapse.

Nuclear contributes to that case by cutting carbon and providing reliable electricity. For some cultures, this aligns with values of stewardship, independence, and responsibility. For others, memories of harm or cultural traditions of reciprocity make it harder to accept.

The moral case is not uniform. It is shaped by culture, history, and identity. But across differences, one truth emerges: *Clean energy is not just technical. It is moral.* It is about how we live together and what legacy we leave behind.

Culture as Compass

Families everywhere use culture as a compass for energy choices. In France, the compass points toward nuclear pride. In Germany, it points toward solar and wind as moral duty. In Japan, it points toward caution born of trauma. In Indigenous traditions, it points toward respect and reciprocity. In the Pacific Islands, it points toward survival.

Culture does not make decisions simple. But it does remind us that energy is never just about megawatts. It is about values. It is about how we explain our choices to our children.

The moral case for clean power — whether nuclear, renewables, or both — rests on the principle every family knows: *We must leave the world better than we found it.*

Family Action:

Talk about what clean energy means in your family's values — fairness, care for creation, or justice for future generations. Learn how your choices reflect those beliefs, and commit to one action that turns your principles into power — like supporting community solar, choosing renewable electricity, or helping a local school cut its energy use.

Part IV

Looking Ahead

The future of energy belongs to the next generation. This part explores education, equity, jobs, fusion, SMRs, and how nuclear can work with renewables.

CHAPTER 19

Energy and Education — Teaching the Next Generation

What children learn today will power tomorrow

Why Energy Belongs in the Classroom

Every family knows that education shapes the future. Parents remind children to finish homework not because the worksheet itself matters, but because it builds habits of thinking, resilience, and imagination. Energy too belongs in this sphere.

For most of the last century, energy was almost invisible in schools. Children learned math, literature, and history, but not where electricity came from or how it shaped their lives. That is changing. Around the world, teachers are weaving energy into lessons not just as science, but as citizenship and ethics.

Because when children learn about energy, they are not just learning about kilowatts. They are learning about fairness, stewardship, and the power of choices — the same choices that will guide nations as they tackle climate change and build cleaner economies. Lessons about the sun and wind become lessons about community and justice. Experiments with circuits become conversations about who has the power — literally and politically.

Science Fairs and Solar Ovens

In Kingston, Ontario, a fifth-grade science fair featured a girl named Samira who built a solar oven out of a pizza box, tinfoil, and black paint. She baked cookies under the winter sun, proudly offering them to classmates. Her project sparked questions: If cookies can bake with sunshine, why not dinner? Why not homes?

Stories like Samira's matter. They show that energy education doesn't have to be abstract. It can be hands-on, tasty, and fun. Parents who watched the experiment came away asking their own questions. Children became teachers.

Similar projects are popping up elsewhere. In rural Colombia, students design pedal-powered blenders to make fruit smoothies while learning about kinetic energy. In Vietnam, middle schoolers build tiny hydro turbines in classroom aquariums to power LED lights. In Jordan, a "Sun School" program invites families to evening festivals where children display solar lanterns and teach neighbours how they work.

These activities turn classrooms into laboratories of imagination. A 2023 UNESCO survey found that students who participate in hands-on energy projects are 40 percent more likely to pursue engineering or environmental science degrees later on. When children see energy in action, they begin to see themselves as participants, not bystanders.

Germany: *Energiewende* in the Classroom

In Germany, energy education is inseparable from culture. After the decision to phase out nuclear and expand renewables, schools began teaching children about wind and solar as part of national identity.

In Hamburg, students visit offshore wind farms as part of field trips. In Berlin, schools encourage students to monitor classroom electricity use, competing to see which group can save the most. Teachers explain that every kilowatt-hour saved is a small act of responsibility.

For German families, children often come home preaching efficiency.

Parents laugh when kids scold them for leaving lights on, but they also feel pride. The federal education ministry reports that these programs cut participating schools' electricity use by nearly 15 percent — proof that learning changes behaviour. Energy education turns values into habits, shaping not just bills but culture.

Even kindergartens have joined in. In Freiburg, preschoolers sing songs about the sun while helping teachers measure rooftop solar output with colourful gauges. From their earliest school days, children connect the warmth on their faces to the electricity in their lights.

Japan: Lessons from Fukushima

In Japan, energy education carries a weightier tone. After the Fukushima disaster, schools wove lessons about safety, risk, and trust into everyday learning. Children are taught how radiation is measured, how evacuation plans work, and why transparent communication matters as much as engineering.

In Sendai, a boy named Hiroshi explained to his classmates how a Geiger counter clicks when it detects radiation. His teacher guided the class into a wider discussion, asking, "What should we do if leaders don't tell the truth?" Another student responded, "Keep checking for ourselves and keep asking questions." What began as a science exercise became a civics lesson in accountability.

Field trips now take students to both renewable-energy sites and carefully monitored nuclear facilities. They meet engineers who answer every question, even the hardest ones about past mistakes. Parents say these visits help children replace nameless fear with understanding.

Universities and high schools also run "citizen science" projects where families collect soil and water samples for independent testing. Grandparents who once lived near the Fukushima exclusion zone join their grandchildren in these projects, turning personal loss into shared inquiry. The Japanese Ministry of Education reports that participation in such programs has doubled science club enrollment since 2015. For families, these activities are a form of healing: They acknowledge trauma, honour memory, and give the next generation tools to demand openness.

Kenya: Solar Scholars

Through Kenya's "solar scholars" program, students bring small solar kits home from school, extending the reach of electricity beyond classrooms. Families that once spent a significant share of their income on kerosene now see children studying under steady light, while schools themselves power fans, charge devices, and keep vaccines cool — linking energy access directly to learning and health.

One student, Amina, proudly showed her parents how to wire a panel to charge their phone. Her father laughed and said, "She is now the engineer of the family." Teachers report that students like Amina routinely outperform peers on national exams, simply because evening study is now possible.

The ripple effects extend well beyond school walls. Mothers use solar lamps for evening weaving cooperatives, boosting household income. Local health clinics keep medicines cool without noisy diesel generators. Village councils have begun discussing small solar microgrids to power grain mills and irrigation pumps.

International partnerships amplify the impact. Kenyan universities pair engineering students with rural schools to improve panel efficiency and battery safety. Families who once thought of electricity as a distant government promise now see it as something their own children can create and manage. Energy education becomes economic education, inspiring small businesses and new career dreams.

Canada: Nuclear Literacy

In Ontario, schools near nuclear plants often organize visitor days where families tour plant facilities, see control rooms, and talk with engineers face-to-face. Children learn that reactors are not mysterious monsters but carefully managed systems protected by layers of safety protocols.

In science centres and museums, this difference is made visible. Interactive exhibits place a tiny uranium pellet next to images of the coal needed to generate the same power. Parents watch as children pause, surprised, trying to reconcile the scale of what they see with the size of what they are holding.

Some high schools now offer elective courses in nuclear science and clean energy policy, often co-taught by working engineers. Students shadow reactor operators or join internships in medical-isotope labs, where they see nuclear technology saving lives through cancer treatments. Teachers note that even students who never considered science careers leave with a clearer sense of how reactors fit into Canada's clean energy goals.

Indigenous communities are also part of the conversation. In Northern Ontario, joint workshops bring together reactor specialists and First Nations leaders so students can hear multiple perspectives on waste storage, land stewardship, and long-term safety. These dialogues teach young people that nuclear literacy is not just physics — it is also culture, consent, and responsibility.

Religion, Ritual, and Storytelling in Education

In many cultures, energy education unfolds not only in classrooms but also in temples, churches, mosques, kitchens, and around village fires. Long before the words "carbon footprint" or "kilowatt" were common, families used faith and story to teach respect for the forces that power life.

Consider the morning prayers of Hindu households along the Ganges. Parents and grandparents rise with the sun and chant verses from the *Rig Veda* praising Surya, the solar deity. Children join in, learning through rhythm and repetition that sunlight is a sacred gift to be honoured, not squandered. When those same children later help place a small solar panel on the roof, the act feels like a continuation of an ancient ritual rather than a break from it.

In Catholic communities across Latin America, Pope Francis's encyclical "Laudato Si'" has become a guidebook for everyday conservation. Families read passages during evening devotions, linking the act of turning off unused lights or sharing a bicycle ride instead of driving to the Pope's call to care for "our common home." Priests bless rooftop solar arrays in village ceremonies, framing them as both environmental stewardship and spiritual gratitude.

Muslim families weave energy consciousness into daily prayer as well. The Qur'anic principle of *khalifa* — humanity's role as caretaker of the

Earth — turns simple acts into moral ones. In Jakarta, an imam begins Friday sermons by reminding parents that wasting electricity or water is a form of *israf*, or excess. Children absorb the idea that conserving resources is a way to honour God, not merely to save money.

Indigenous traditions provide some of the most vivid examples of energy literacy through story. In many First Nations communities of Canada, elders tell the tale of how the rivers chose to share their power with people only if they promised to give thanks. Around campfires, children learn that each fish taken and each log burned carries responsibility. In the American Southwest, Navajo families recount the painful history of uranium mining on their land while also teaching ceremonies of respect for the sun and wind, ensuring that new energy projects begin with community consent and prayer.

In West Africa, the Yoruba people celebrate the annual Oshun festival, honouring the river goddess who sustains crops and communities. Parents explain that the river's health — and by extension the power of local hydro dams — depends on careful stewardship. Songs and drumming become lessons in water conservation, passed from elders to children.

Even seemingly secular cultures carry echoes of this spiritual storytelling. In Finland, families observe *Kaamos*, the long winter darkness, by lighting candles and sharing folk-tales of the sun's return. These evenings become informal classes in energy awareness, where conserving precious firewood or electricity is part of honouring nature's rhythms.

What unites these diverse practices is the way they blend emotion, spirituality, and practicality. A verse recited at dawn, a blessing of a solar panel, a grandmother's fireside story about the river's gift — all teach children that energy is not only mechanical but moral. They learn that every watt and flame carries history and responsibility.

Modern educators increasingly recognize the power of these traditions. Environmental curricula in places like New Zealand now invite Māori elders to share legends of Papatuanuku (Earth Mother) alongside lessons on geothermal energy. In Alaska, Yupik storytellers open science classes with tales of the northern lights, connecting renewable wind projects to ancestral respect for the sky.

By linking faith, ritual, and story to modern energy choices, families create an emotional foundation that outlasts textbooks. When a child raised with these traditions grows up to vote on climate policy or design a solar microgrid, the decision is guided not just by numbers, but by values learned on prayer mats, at festival altars, and beside crackling fires.

Family Habits as Education

Not all education comes from schools. Families teach energy literacy through daily habits that quietly shape a child's worldview. Parents involve children in reading electricity bills, explaining how each appliance contributes to the total. Teens help install smart thermostats and learn to program schedules that save both energy and money. Siblings compete to see who can shorten their showers or remember to turn off lights, turning conservation into a game.

In Mexico City, one family posts a weekly "energy scoreboard" on the fridge, rewarding the winner with extra dessert or a later bedtime. In Finland, grandparents teach grandchildren to dry laundry on outdoor lines instead of running the dryer, sharing stories of wartime frugality and the smell of sun-dried sheets.

Families in the United States have adopted "energy sabbath" evenings — one night a week lit only by candles or solar lanterns — using the quiet time for storytelling and board games. Parents say the tradition sparks conversations about what life was like before electricity and why thoughtful consumption matters.

Research from the International Energy Agency shows that households that involve children in energy-saving decisions cut consumption up to 12 percent more than those that don't. Children raised with these habits often carry them into adulthood, shaping how they vote, how they build their own homes, and even how they choose careers. Energy wisdom becomes a family inheritance, passed down as naturally as recipes or holiday traditions.

Teaching for Tomorrow

The energy transition is not only technical. It is cultural and educational. The choices families make tomorrow will be shaped by what children learn today.

In France, children learn pride in nuclear independence. In Germany, they learn identity through solar and wind. In Japan, they learn caution and vigilance. In Indigenous communities, they learn respect and reciprocity. In the Pacific Islands, they learn urgency and survival. And increasingly in places like India, Brazil, and the Middle East, national curricula are embedding climate science and clean energy modules so every graduate understands both opportunity and responsibility.

New partnerships are springing up: universities mentoring high school robotics teams on building small wind turbines, city libraries lending out home energy meters alongside books, youth exchanges where students from coal regions visit offshore wind farms and vice versa. These programs make the abstract transition tangible, and they remind parents that education is not a spectator sport.

Families everywhere should see energy as part of education. Because in the end, teaching children about kilowatts is not about watts at all. It is about values, habits, and imagination. It is about preparing them not just to inherit the energy system but to shape it.

Every parent knows that children remember stories more than lectures. The story we tell about energy — of pride, caution, respect, survival — will become the compass children use to guide the future. And the classrooms, kitchens, and living rooms where those stories are told are as vital to the clean energy transition as any solar panel or reactor.

Family Action:

Encourage energy learning by borrowing kits from libraries or schools and setting up simple experiments that connect to your local grid story.

CHAPTER 20

Just Transitions and Equity

Leaving no family behind in the energy shift

The Unease of Change

Every family knows that change, even for the better, can be unsettling. Moving houses means new schools for children, new commutes for parents, and new routines for everyone. The promise of a better home does not erase the pain of leaving the old one.

The global energy transition is no different. As societies move from coal, oil, and gas toward renewables and nuclear, millions of families will be affected. Some will gain new opportunities. Others fear losing everything. Economists estimate that more than *thirty million workers* worldwide are tied directly to fossil-fuel supply chains, and tens of millions more depend on related industries — rail lines that haul coal, ports that ship oil, restaurants that serve mining towns.

A just transition means ensuring that no family is abandoned in the dark. It means fairness, dignity, and support for those who gave decades to building the old energy system, even as we shift to the new. The choices governments make in the next decade will decide whether those workers become casualties or pioneers of a cleaner era.

Appalachia: When Coal Declines

In the hills of West Virginia, ten-year-old Sarah once believed her father's job as a miner would last forever. Coal was identity, heritage, and livelihood. For generations, the mine had been more than a workplace — it was a rite of passage. Sarah's grandfather kept a framed black-and-white photo of his first day underground, and family stories at Sunday dinners often began with the words, "Back when I was working the seam . . ."

But as mines closed, Sarah's father faced unemployment. Bills piled up, neighbours moved away, and the town's high school graduating class shrank by half. For Sarah's family, the transition felt less like opportunity and more like betrayal. The local grocery store shut its doors after ninety years; the Friday night football team had to merge with a rival county just to have enough players. The cultural heart of the community seemed to flicker along with the mine lights.

Yet across the region, *new programs are quietly transforming despair into possibility.* Former miners are being trained to install solar panels, maintain wind farms, and refurbish nuclear reactors. Community colleges now offer "mine-to-solar" certificates, and union halls run evening classes on electrical work and industrial robotics. A partnership between a regional university and a national lab teaches laid-off workers how to operate advanced battery-storage systems, while a women-led nonprofit offers scholarships for daughters of miners who want to study engineering.

Families are beginning to imagine a future that preserves dignity even as the economy shifts. One local pastor now hosts "career potlucks," where parents share homemade meals while listening to graduates of solar-training programs describe their new jobs. Another town converted its shuttered coal depot into a makerspace with 3D printers and welding stations so residents can learn modern manufacturing skills.

Hope, the families say, doesn't come from political slogans or one-time aid checks. It comes from *real jobs and shared pride* — work that keeps roofs overhead and traditions of craftsmanship alive. As one father told his children while they watched workers raise the first turbine on a reclaimed mountaintop: "Your grandfather dug the coal, I worked the shaft, and maybe you'll keep these blades spinning. Different tools, same strength."

Germany: Phasing Out Coal with Compensation

In Germany's Ruhr Valley, coal was once king. Generations of families worked in mines, building communities around steady paychecks. When Germany pledged to phase out coal, families feared devastation. Grandparents who had watched the mines provide security for a century wondered if their grandchildren would have to leave home to find work.

But the government responded with a *deliberate package of fairness*: billions of euros for affected regions, retraining programs for workers, investments in renewables, and cultural preservation. Former miners now help build wind turbines or maintain solar plants. Museums and cultural centres celebrate the mining past even as children tour new hydrogen pilot plants, learning that history and progress can share the same streets.

The transition is not painless, but it is softened by dialogue and compensation. Parents can tell children, "Your grandfather dug coal, your father built turbines, and you may code the software that runs them. Each of us has a role." In community halls, miners' choirs still sing the old songs, but now at festivals powered by solar arrays.

South Africa: Blackouts and Hope

In Johannesburg, families endure rolling blackouts. Children study by candlelight, food spoils in fridges, and small businesses lose income. Coal dominates the grid, but it is aging, unreliable, and polluting. Families pay the price.

South Africa's government is pursuing a *Just Energy Transition*, with billions pledged for renewables, jobs, and community support. Pilot programs in Mpumalanga now pay former coal workers during training so families can keep food on the table while learning new trades. Local colleges teach solar installation and wind-farm maintenance; some even pair trainees with German apprentices to share expertise.

The stakes are high. For South African families, a just transition means the difference between darkness and dignity. Parents like Nomsa Khumalo tell their children that change is coming — "But this time," she says, "you will have a say in how it happens."

Canada: Nuclear Jobs and Equity

In Ontario, *massive nuclear refurbishment projects* along Lake Huron and Lake Ontario have created thousands of skilled trades jobs. These are not just short-term contracts but decades-long opportunities for welders, electricians, machinists, and software engineers. Families benefit from stable employment, good wages, and community investment. Towns that once worried about plant closures now buzz with new cafés, childcare centres, and training institutes.

Parents tell their children that nuclear work is not only safe but secure. "The plant kept our lights on through every ice storm," one mother says. "Now it's helping you pay for college."

But the question of *equity* looms large. Will women, Indigenous youth, and racial minorities share in these opportunities? Historically underrepresented groups are now being targeted with scholarships, apprenticeships, and mentorships. Partnerships with First Nations bring Indigenous students into engineering co-ops, while women's trade networks run mentorship programs to break old barriers.

Some communities are *rewriting the rules of participation*. The Saugeen Ojibway Nation negotiated agreements to ensure Indigenous voices guide environmental monitoring and to guarantee a share of long-term economic benefits. Colleges near the Bruce and Darlington plants now run "bridge programs" for newcomers to Canada who were engineers or technicians in their home countries, helping them earn local credentials so their skills are not wasted.

Families view these programs as steps toward reconciliation. Grandmothers who once protested uranium projects now support their granddaughters' chosen apprenticeships, backed by community consent and oversight. Parents proudly describe nuclear careers as *a bridge between old injustices and new inclusion*, a chance to earn a living while contributing to clean air and climate goals.

India: Solar Dreams

In rural India, families once relied on kerosene lamps and diesel pumps. Today, solar farms are spreading across the countryside. Youth once destined for farm labour are now learning to install and maintain panels. For parents, this means hope: Their children may have brighter futures than they did.

Yet justice requires inclusion. Not every region has equal access. Some families worry that wealthier states will capture most of the benefits. Grassroots cooperatives are beginning to change that, pooling microloans so even the poorest villages can finance panels and training. Women's groups in Rajasthan now own shares in local solar arrays, and community radio stations broadcast lessons on maintaining batteries and wiring. A just transition means ensuring that opportunity reaches every household, not just those near cities or in politically favoured districts.

Australia: Mining and Identity

In Queensland and New South Wales, coal mining built communities. Families feel pride and loyalty to industries that put food on the table. As global demand for coal declines, parents ask what comes next. Some welcome new jobs in renewables or nuclear. Others feel abandoned, fearing their children will have to leave to find work.

Australia's challenge is cultural as well as economic: helping families see that identity can shift without erasure, and that pride can move from coal to clean energy without dishonouring the past. Local councils now sponsor storytelling projects where retired miners share histories alongside young solar technicians, showing that heritage can evolve rather than vanish. In Broken Hill, school projects pair students with elders to record mining memories and design solar sculptures, blending old pride with new possibility.

The Dimensions of Equity

Just transitions are not only about jobs. They are about *fairness across every axis of life*: gender, race, class, geography, and generation.

Women, long underrepresented in energy, must be welcomed into trades, engineering, and leadership. In too many countries, less than 25 percent of energy-sector workers are female, and even fewer hold decision-making roles. Equity means scholarships for girls in STEM, paid apprenticeships that include childcare, and safe worksites where harassment is not tolerated.

Families of colour, often excluded from energy careers, need access to training and capital. In the United States, Black households are more likely to live in older, drafty homes that waste energy and drive up bills. Programs that retrofit these houses — insulating walls, replacing furnaces, installing heat pumps — become acts of climate justice as much as energy efficiency.

Indigenous families, whose lands hosted past projects, must share ownership of future ones. In Canada and Australia, legal agreements now give Indigenous nations revenue shares and veto power over major energy developments. In Kenya and Tanzania, Maasai leaders are co-designing wind projects to preserve grazing routes while supplying electricity.

Equity also stretches beyond borders. Families in low-income nations contributed least to the climate crisis yet face the worst floods, droughts, and storms. Global climate funds that help them build solar farms or seawalls are not charity; they are *a moral debt payment.*

And equity crosses time. Children and grandchildren will live longest with today's decisions. A fair system protects not just today's ratepayers but tomorrow's citizens, ensuring that the next generation inherits a livable planet instead of an unpaid bill of damages.

Analogies of Transition Families Know

Parents often explain complicated transitions through the *language of daily life*. Moving houses means packing memories but opening new doors. Changing schools means saying goodbye to friends but meeting new ones. Shifting careers means carrying old skills into new contexts.

The energy transition works the same way. Families carry pride in the past — coal, oil, gas — but step into new futures of solar, wind, nuclear, and storage. The skills remain valuable, the dignity intact, the identity respected.

A former miner in Kentucky puts it simply to his kids: "I still work with power; it just comes from the sun now." In Germany, a grandmother tells her grandchildren, "Our coal heated Europe, and now our wind will light it. Both are our legacy." A South African father compares the shift to a family recipe: "We still cook the same meal of community and hard work, but the ingredients have changed."

Such analogies matter because they *bridge emotion and reason.* They reassure children — and parents — that letting go of one way of life does not mean abandoning history or worth. The story of energy becomes the story of growth: carrying forward courage, skill, and pride while embracing cleaner tools for the sake of those yet to come.

Fairness in the Future

The transition to clean energy is inevitable. The question is whether it will be just. Families must not be left behind. Communities must not be abandoned. Equity must be more than a slogan.

A just transition means retraining miners in Appalachia, compensating families in Germany, creating solar jobs in India, building equity in Canada, and restoring trust in South Africa. It means telling children that the future will be fair — not only clean, but just.

Because in the end, the energy transition is not only about technology. It is about people and the promises societies make to them. Justice means ensuring that *every family has light, dignity, and a place in the new energy world.* Only then will the shift to clean power be remembered not just as an environmental triumph, but as a moral one.

Family Action:

- Talk about who in your community might be left behind in the energy shift — retired workers, low-income families, or Indigenous neighbours — and find one way to help bridge that gap.
- Support local retraining programs, share resources, or amplify stories of fairness and inclusion. A just transition begins with just families.

CHAPTER 21

Pathways into the Industry

How families can step into the clean energy future

From Consumers to Builders

For decades, families were told their role in energy was simple: Pay bills, flip switches, and turn off lights when not needed. The system was something distant — planned by experts, managed by utility companies, and delivered silently to homes.

But the clean energy transition is rewriting that script. Families are no longer just consumers; they are builders. Their children are tomorrow's welders, electricians, engineers, coders, and policymakers. The energy shift will be shaped not only in government chambers but in workshops, classrooms, and everyday family decisions about careers and education.

The International Labour Organization now estimates that clean energy growth could create more than *fourteen million net new jobs worldwide by 2030.* That figure is discussed around dinner tables from Ontario to Odisha. For parents, it is an invitation to encourage energy careers as they once encouraged medicine or law. For teenagers, it is a chance to join an industry that feels as essential as it is inspiring — a field that can cut carbon, strengthen communities, and pay the bills.

A Welder in Ontario

When Ontario shuttered its last coal plants, thousands of fossil-fuel jobs disappeared. For families in towns built around those plants, the future once looked uncertain.

Michael, a welder from Kincardine, remembers the worry. "We thought the next generation would have to move away," he recalls. But when massive nuclear refurbishment projects began at Bruce and Darlington, new trades jobs opened. Michael retrained to work on reactor components and pressure vessels. His daughter says with pride, "Dad's work keeps our town alive. He tells us he's part of building clean air."

Colleges partnered with utility companies to create *dual-credit programs*, allowing high school seniors to graduate with both a diploma and an industry certificate. Local unions added night classes so that parents like Michael could learn new techniques without leaving their current jobs. These programs turned what could have been economic decline into a multigenerational story of renewal. Nuclear jobs have become a kind of family inheritance — skills and steady paychecks passed from parent to child like a trade or a craft.

A Student in South Africa

In Johannesburg, rolling blackouts are an everyday frustration. Yet in Soweto, eighteen-year-old Thandi enrolled in a solar-installation course run by a community training centre. Her mother beams: "She will bring light not just to our house, but to our whole neighbourhood."

These programs do more than teach wiring or panel placement. They provide *stipends and childcare*, so students can keep supporting their families while they learn. Graduates leave with both certification and guaranteed apprenticeships on regional solar farms. For families like Thandi's, an energy career is more than a technical job — it is nation-building. Each installation reduces the need for coal, and each paycheck strengthens local resilience against the blackouts that have long defined daily life.

Germany's Green Apprenticeships

In Germany, apprenticeships are a proud tradition, blending classroom instruction with hands-on training. The country's *Energiewende* — its ambitious energy transition — has given these programs new life.

Jonas, a twenty-year-old from Hamburg, climbs turbines on the North Sea to maintain blades taller than church steeples. "We never thought of wind as a career," his parents admit, "but now it's his identity." Municipal utility companies sponsor *family days* at wind farms, so parents and grandparents can see the technology firsthand and understand the safety measures their children work with every day. These gatherings turn abstract climate goals into local pride and create new narratives of stable, respected work.

Indigenous Youth in Canada

For decades, Indigenous communities were excluded from the profits and decision-making of energy projects built on their lands. That is beginning to change.

In Northern Ontario, Aiden, a young member of the Anishinaabe Nation, joined a nuclear apprenticeship program at a nearby plant. His grandmother remembers the scars of past uranium mining but supports his choice: "This time, we are at the table."

Modern agreements include *revenue sharing, cultural mentorship, and joint governance boards*, ensuring that when Aiden's plant produces power, his community shares both the income and the authority. These partnerships mean that Indigenous youth are not merely workers; they are co-owners and future leaders, blending technical expertise with ancestral stewardship.

India's Solar Surge

In Rajasthan, vast fields once known only for dust and hardy crops now gleam with mirrors and panels. Young men and women who once migrated for seasonal labour now work year-round as solar technicians.

Parents marvel: "The sun has always been with us. Now our children earn their living from it." Women's cooperatives have launched all-female installation teams, challenging long-entrenched gender roles. Training centres offer literacy classes alongside technical instruction so that even adults who never finished school can qualify for maintenance jobs. For many families, solar careers mean children stay close to home for the first time in generations, keeping cultural ties strong while building economic security.

Veterans and Nuclear in the United States

For many United States service members, careers in civilian nuclear power feel less like a career change and more like a natural continuation of duty. Navy sailors who once kept submarine reactors humming often say the hum of a civilian plant's turbine feels like the heartbeat of home.

When Marcus left the navy after a decade aboard a nuclear-powered carrier, he wondered how to translate his skills into civilian life. A recruiter from a Midwestern utility called within weeks. The company's training program recognized his navy license, allowing him to skip months of preliminary coursework. "I went from defending the nation to powering it," he explains. His children, who grew up hearing stories of long deployments, now visit the plant on family days and proudly introduce their father as "the man who keeps the hospital lights on."

Utility companies across the United States actively court veterans like Marcus. Georgia's Vogtle plant and Tennessee's Watts Bar station run "military-to-civilian" transition programs that provide housing stipends and mentorship for spouses seeking local employment. Some states — including Illinois and South Carolina — offer tax credits or tuition reimbursement for veterans entering nuclear operations or radiation protection programs.

Military families often speak of the cultural continuity: strict safety procedures, a team ethic, and an understanding that a small mistake can have large consequences. One navy veteran put it this way: "In the service, I protected sailors. Now I protect communities. The mission hasn't changed, only the scenery." For children in these households, nuclear work becomes a symbol of stability after years of frequent moves — a career where discipline and patriotism light homes instead of aircraft carriers.

Expanding Equity and Inclusion

Clean energy growth promises millions of jobs, but opportunity is not yet distributed evenly. Women still make up less than a third of the global renewable workforce and an even smaller share of nuclear engineers and senior plant operators. Families of colour and recent immigrants often face barriers — high tuition, lack of professional networks, or simply the absence of visible role models. Rural youth can struggle to reach training centres or even access the broadband needed for online coursework.

Communities are working to change that. Women in Nuclear chapters in the United States, Canada, and Europe pair high school girls with plant engineers for job-shadow days and long-term mentorship. In Texas, the Mujeres de Energía program provides bilingual welding and electrical courses so Spanish-speaking mothers can enter the booming wind industry. Indigenous Clean Energy in Canada funds apprenticeships that include cultural mentorship, ensuring young workers can practice traditional ceremonies while pursuing technical careers.

Cities are investing too. New York's Energy Equity Academy offers stipends and childcare assistance to low-income parents while they train for solar-installation jobs. In Kenya and South Africa, mobile training labs travel to rural villages so that students — especially young women — can learn turbine maintenance without leaving their families for months at a time.

Representation matters not only for fairness but for the strength of the workforce itself. Research from the International Renewable Energy Agency shows that diverse teams are more innovative and have lower turnover. For families, these efforts mean a daughter can imagine herself as a nuclear safety analyst, an immigrant son can see a future as a wind-farm engineer, and grandparents can watch their grandchildren thrive in industries that once seemed closed to them.

Analogies Families Understand

Parents often explain career transitions with images that children can feel in their bones. A carpenter guiding a small hand along a smooth piece of oak shows how patience shapes strength. A farmer teaches children to plant seeds and wait for the quiet magic of growth. A cook invites the next generation to taste, stir, and finally own the family recipe. Each lesson carries more than technique — it carries identity.

Energy careers follow the same rhythm. A grandfather in Kentucky who once descended coal shafts now watches his granddaughter climb turbine towers. Around the kitchen table, they trade stories: He speaks of the rumble of mining carts; she describes the whistle of mountain winds against spinning blades. Different sounds, same pride.

In the Navajo Nation, elders who once herded sheep near uranium mines now bless new solar farms built with tribal ownership. They tell children that the "power of the sun" is the same force that warmed their ancestors, only now it's captured with panels instead of prayers alone. In rural Germany, a retired machinist teaches his grandson how to keep wind-turbine gears clean, recalling the exacting precision he once used on coal-plant turbines.

These analogies remind families that technology may change, but dignity does not. Passing down an energy career is like handing over a well-worn set of tools or a favourite recipe: It's an inheritance of skill, discipline, and purpose. When children grasp that continuity, the transition from fossil fuels to clean power stops feeling like a break with the past and starts to feel like the natural next chapter of a family story.

Building Futures Together

Pathways into the industry are not abstract pipelines; they are real routes to security and purpose.

In Ontario, welders find new futures in nuclear refurbishment. In South Africa, solar training gives daughters dignity and communities light. In Germany, apprentices climb turbines with pride. In Canada, Indigenous youth step into leadership roles once closed to them. In India, solar fields keep children close to home. In the United States, veterans continue a lifelong mission of service through civilian nuclear power.

The clean energy transition is ultimately a human project. Families everywhere deserve to see themselves not just as consumers of electricity but as participants, builders, and leaders. Policies may open the door, but it is families — one apprenticeship, one retraining program, one dream at a time — who will walk through and build the future.

Family Action:

- Visit local clean energy sites, from solar fields to nuclear facilities, to see how real people are building the future — and discuss what roles your family might play.
- Partner with schools or community groups to host "Energy Futures" nights where families can meet local technicians, engineers, and apprentices to learn firsthand about career paths.

CHAPTER 22

Fission and Fusion — Two Paths to Power

Splitting and joining atoms

Why Families Need Both Stories

When families talk about nuclear energy, the conversation often splits in two directions. On one path lies *fission*, the process of splitting atoms, which already powers homes and hospitals across the world today. On the other is *fusion*, the dream of joining atoms — the same reaction that fuels the sun — still just beyond our reach but captivating our imagination.

Fission is the dependable present, keeping refrigerators cold in Ontario, illuminating Parisian boulevards, and running subway trains in Seoul. Fusion is the dream of tomorrow, promising virtually limitless, clean energy if only we can unlock it. Both paths matter, and families deserve to understand them not as abstractions but as living stories shaping the world their children will inherit.

Parents who talk about these twin narratives are really teaching more than physics. They are exploring ethics, responsibility, and imagination: how we balance today's needs with tomorrow's possibilities, how science and citizenship intertwine, and how careful choices today — funding research, demanding transparency — set the stage for discoveries their grandchildren may witness.

Fission: The Fire We Already Hold

Fission is the workhorse of nuclear power. At its heart, it is simple to explain. Imagine a Lego tower: When it is knocked apart, pieces scatter with energy. Atoms are like that. Splitting uranium atoms releases heat, which boils water into steam, spins turbines, and generates electricity.

For families, the outcome is familiar. The electricity from fission powers the coffee maker in the morning, the refrigerator in the kitchen, and the hospital ventilator down the street. Ontario offers a clear example. Because of its nuclear fleet, coal was eliminated entirely. Parents noticed fewer smog days, children with asthma breathed easier, and families enjoyed cleaner skies.

In France, nuclear identity runs deep. Since the 1970s oil crisis, fission has provided independence and stability. Families there live with steady bills and one of the cleanest grids in the developed world. South Korea treats fission as both practical and cultural — a symbol of progress that keeps its high-tech economy humming.

Other nations show fission's versatility. In the United Arab Emirates, the new Barakah plant powers desert cities and provides fresh water through desalination, turning seawater into drinking water for millions. In Finland, the Olkiluoto 3 reactor recently came online, giving families long-term energy security during volatile European gas markets. Even tiny island nations such as Palau are now studying compact reactor designs to reduce costly diesel imports.

Yet fission also stirs unease. Parents remember Chernobyl and Fukushima. They ask about safety and waste, worrying about what their grandchildren might inherit. These concerns are real, yet so are the benefits. Modern reactors include multiple layers of protection, passive cooling systems, and emergency protocols as routine as fire drills in schools. Independent regulators publish real-time radiation data online, and many plants invite families to open houses so they can see safety procedures firsthand.

For families weighing climate risks, fission remains one of the fastest, most scalable ways to cut carbon today — an everyday shield against a warming world. It is, as one French engineer told a group of visiting students, "the quiet giant in the fight against climate change."

Fusion: The Dream of Bottling a Star

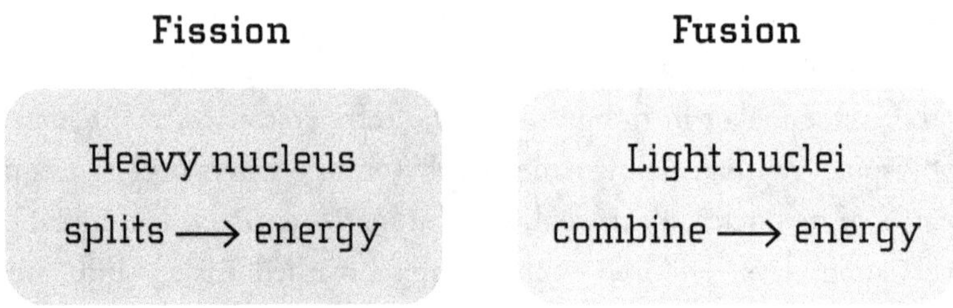

Fig. 17. Fission vs fusion simplified

If fission is the firewood split to make a campfire, fusion is the sun itself. Instead of splitting atoms, fusion joins them together. When two hydrogen atoms fuse, they release enormous energy. It is the reaction that makes the sun shine, and it could one day power humanity.

Scientists have been trying to recreate this reaction for decades. The challenge is immense: Fusion requires temperatures hotter than the sun's core, held in place by powerful magnets or lasers. It is like trying to bottle a star.

Still, families are inspired by the pursuit. In southern France, the international ITER project brings together dozens of countries to build the largest fusion experiment in the world. Local families speak with pride about hosting a dream for the planet, knowing their children might witness history. In the United States, headlines about the National Ignition Facility's brief but historic "net energy gain" shots make parents and children alike ask: "Could this end the age of fossil fuels?" In Japan, the JT-60SA tokamak is a point of quiet national pride, while in the United Kingdom and Canada, private startups race to turn lab successes into commercial reactors.

Fusion has always seemed just over the horizon — often jokingly referred to as being "thirty years away." Yet each year progress inches forward. Magnetic-confinement machines now hold plasma for minutes instead of seconds. Advanced lasers squeeze hydrogen targets with ever greater precision. Entire university programs are emerging to train the first generation of fusion engineers, and philanthropic funds are supporting public—private partnerships.

For families, fusion remains a promise worth chasing, even if the wait is long. Like teaching a child to play piano, progress can feel slow — yet each small improvement builds toward a transformative moment. Parents who follow the news explain to children that breakthroughs come not in leaps but in thousands of careful steps.

Fission and fusion are not rivals; they are chapters in the same unfolding story of human ingenuity. One provides the strength of a well-built house, the other the promise of a vast horizon. Fission is here, already cutting carbon and running our cities day and night. Fusion is coming — and its pursuit has already united scientists, governments, and entrepreneurs from countries that rarely collaborate on anything else.

For families, the relationship between the two can be imagined in everyday terms. Think of a household budget: A reliable salary covers today's groceries and bills, while a long-term investment account grows quietly for tomorrow's college fund. Fission is that dependable paycheck, steady and predictable. Fusion is the investment — not guaranteed, but filled with transformative potential. Together, they make a family secure.

Energy planners increasingly design for this kind of complementarity. Detailed scenarios from the International Energy Agency and MIT's Energy Initiative picture grids where large fission plants provide stable baseload electricity while fusion reactors, once commercial, deliver flexible bursts of power to handle sudden spikes in demand, desalinate seawater, and generate clean hydrogen for heavy industry. In these visions, solar and wind still play crucial roles, but nuclear's twin forms give the backbone and muscle that weather-dependent renewables sometimes lack.

Families can picture it another way: Imagine preparing a holiday meal. The soup and bread are ready now — nourishing, reliable, and enough to keep everyone happy. In the kitchen, a roast slowly cooks, filling the house with aroma and anticipation. No one would cancel dinner because the roast isn't finished; they simply enjoy what's ready and look forward to what's coming. Fission feeds us today, while fusion fills the air with the scent of tomorrow's possibilities.

Parents who share these analogies help children see that the energy future is not a zero-sum contest. It is a partnership in which today's dependable power supports the patient work of building something even greater.

Families Living the Story

Around the world, families are already shaped by this dual path. In Ontario, they breathe cleaner air thanks to fission. In France, they enjoy stable bills and pride in independence. In the United States, they follow news of fusion experiments with curiosity and hope. In China and South Korea, nuclear of both kinds is tied to national strength. In developing countries, parents ask whether these technologies will ever reach them, or whether their families will be left behind while others harness the atom.

At dinner tables, parents explain these differences in simple terms. "Fission is like breaking sticks for a fire," one says. "It gives us heat now." Another adds: "Fusion is like building a little sun in a bottle. It is harder, but it could give us light forever." Children understand these stories more easily than numbers or technical charts. And once they understand, they begin to imagine a future where both fire and star can work together.

Family conversations already reflect this. In Boston, a high school science club hosts "fusion Fridays," where parents join students to watch livestreams from international labs. In Nairobi, university students bring younger siblings to public lectures about Small Modular Reactors. Across cultures, the discussion is shifting from *whether* nuclear belongs to *how* these twin technologies can serve everyone fairly.

Walking Both Paths

The climate emergency leaves little room for either-or thinking. Families cannot afford to choose between fighting today's emissions and dreaming of tomorrow's breakthroughs. They must walk both paths at once — holding in one hand the proven reliability of fission and in the other the shimmering promise of fusion.

Walking both paths means practical action. It means supporting policies that keep existing fission plants running safely, upgrading them where needed, and training a new generation of operators and engineers. It also means sustaining the long, meticulous research that will make fusion a commercial reality, funding international collaborations like ITER and smaller private startups that might spring surprises. Governments must provide resources, but so must the public, through patience and the steady insistence that clean power is worth the investment.

For families, this dual commitment can feel familiar. Parents often juggle immediate responsibilities — paying the mortgage, packing school lunches — with long-term hopes like saving for college or planting a tree that will shade grandchildren. They know that tending to the present does not weaken the future; it protects it. Fission is like that mortgage payment, essential every month. Fusion is the tree, growing slowly but destined to outlive those who planted it.

Communities around the world are already modelling this mindset. In Finland, families welcome the steady electricity of fission while their universities host fusion research centres. In the United States, parents who work at existing reactors also mentor students in plasma physics clubs. In South Korea, children visit both operating fission plants and cutting-edge fusion labs, understanding that their country's strength lies in nurturing both.

To walk both paths is also to demand fairness. Families must insist that fission's benefits and fusion's eventual riches are shared equitably — that safety standards are rigorous, that waste is managed responsibly, and that future fusion plants do not simply repeat the mistakes of the fossil era by concentrating wealth and power in a few hands.

Energy, after all, is not only about electrons or engineering. It is about the legacy parents leave when they light a lamp for their children.

Choosing fission for today and fusion for tomorrow is not indecision; it is the deepest kind of care — a promise that the fire we hold now and the star we chase for the future will together keep every family warm, safe, and hopeful.

Family Action:
Follow fusion news by noting which milestones show real progress, like sustained energy gain and practical engineering advances.

CHAPTER 23

Small Modular Reactors (SMRs)

Why small might be the next big thing.

A New Kind of Nuclear Story

Across the world, the conversation about nuclear energy is changing. For decades the word "reactor" conjured images of sprawling power stations: concrete domes, endless security fences, and decades-long construction projects. Today, a quieter revolution is unfolding. Engineers are designing Small Modular Reactors, or SMRs — compact power plants that can be built in sections, shipped by truck or barge, and installed where energy is needed most.

Families are beginning to hear about SMRs in the news, at town meetings, and even around dinner tables. The promise is simple yet profound: *clean, steady electricity delivered by units small enough to serve a single community*, quick enough to build before climate change races further ahead, and flexible enough to fit places where traditional reactors never could.

For parents and children alike, understanding SMRs is not just a lesson in science. It is a story about how technology, trust, and community can reshape the very meaning of power.

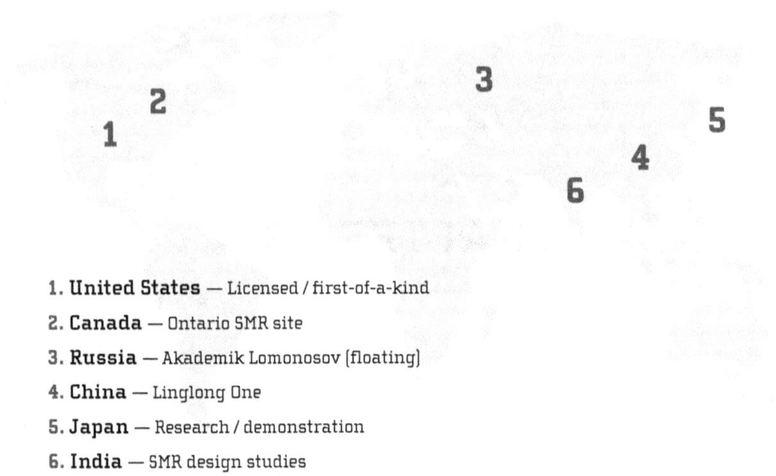

Fig. 18. Selected Small Modular Reactor (SMR) projects worldwide

What Makes an SMR "Small" and "Modular"

The name itself carries two clues. "Small" means the power output is far lower than a conventional gigawatt-scale nuclear plant — typically a few dozen to a few hundred megawatts, enough for a mid-sized city or a cluster of industries. "Modular" means the reactor is built in factory-made pieces, assembled on-site like giant building blocks.

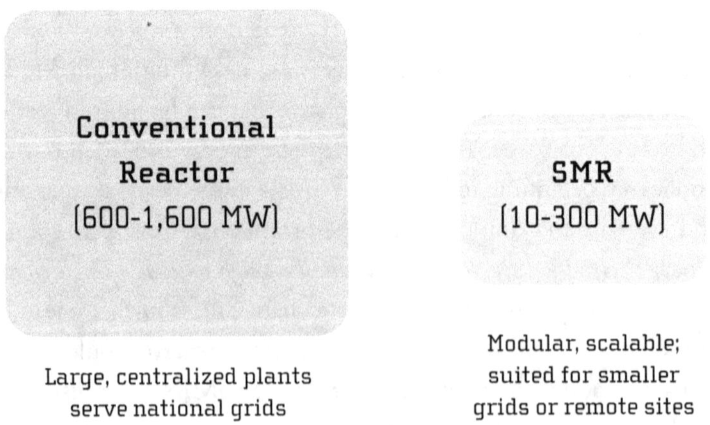

Fig. 19. Scale comparison: Conventional reactors vs SMRs

This approach shortens construction from a decade to a few years, reduces costs, and allows communities to add units gradually, matching demand. A town can start with one reactor and later add another, the way a family adds rooms to a house as it grows. Safety systems are also simplified and often passive, relying on gravity, natural circulation, and physics itself rather than complex pumps or constant operator action.

For families wondering if "smaller" means "less safe," the answer lies in design. Many SMRs are built underground or in sealed steel vessels. Their compact cores generate less heat, and their emergency cooling systems work without electricity. It is like comparing a well-insulated woodstove to a roaring bonfire: The stove is easier to control, even if the fire inside burns just as hot.

Why SMRs Matter for Families

Parents everywhere face two intertwined worries: the rising cost of energy and the rising toll of climate change. SMRs speak to both. Because they can be built where power is needed — near a growing city, an isolated mining operation, or a remote northern village — they reduce dependence on long transmission lines and volatile fuel markets. And like all nuclear power, *SMRs produce virtually no carbon emissions*, helping slow the warming that brings stronger storms, heavier smoke, and higher grocery bills.

For families in diesel-dependent Arctic towns, an SMR could mean heat and light without the weekly roar of fuel trucks. For families in countries with fragile grids, a cluster of SMRs could keep hospitals running when storms topple power lines. And for families everywhere, SMRs offer something harder to measure but deeply felt: the chance to pass on a planet with cleaner air and a steadier climate.

From Drawing Board to Dinner Table

The idea of smaller reactors is not brand new, but only recently have the pieces come together: advanced fuels that tolerate high temperatures, new alloys that resist corrosion, and modular construction methods proven in other industries.

Projects are underway across the globe. Canada is building a first-of-its-kind unit at Darlington, with Indigenous nations involved from the start as co-owners and advisors. In the United States, an SMR design has cleared rigorous federal review, with Utah municipalities planning the first fleet. In the United Kingdom, a national program pairs government funding with private engineering to create factory-built reactors that can be exported like aircraft engines. From Poland to Kenya to Japan, pilot projects are in planning or early construction.

Each project carries its own rhythm, but families notice common themes: *shorter timelines, open communication, and local participation.* The long wait and distant decision-making that once defined nuclear power are giving way to community meetings, transparent safety drills, and opportunities for local hiring.

Living Beside a Small Reactor

What does daily life look like when an SMR comes to town?

Picture a small city that once relied on a coal plant. When the plant closed, jobs vanished and storefronts emptied. Now, on the same riverside site, a compact reactor is rising. At first, residents worry. Will it loom over their streets? Will sirens sound at all hours? But when construction ends, the building is modest and quiet, tucked behind landscaped berms. Children ride bikes past the fence on their way to school. The café across the street sells "Pocket Sun" muffins in a nod to the town's new power source.

Within a few years, tax revenue funds a new library and improved playgrounds. Parents notice the absence of soot on window sills, the cleaner smell of the air after rain. The most surprising thing, many say, is how ordinary the reactor feels — less intrusive than the old coal stacks, more like a steady neighbour than a mysterious giant.

From Skepticism to Stewardship

Communities rarely embrace big projects overnight. Early town halls can be tense, with parents asking about radiation, evacuation plans, and long-term waste. Successful SMR developers have learned that *trust is as essential as steel or concrete.*

In Idaho, engineers invited residents to observe emergency simulations and practice drills before ground was broken. In Canada, Indigenous elders held joint ceremonies with project managers to bless the land and set cultural protocols. Over time, suspicion softened. Residents began referring to the facility as "our reactor," and children drew pictures of it for school science fairs.

These stories reveal a simple truth: *Public engagement is not a box to check but the foundation of acceptance.* Families must feel that the reactor is theirs — something built with them, not for them.

Shared Ownership and Local Pride

An emerging trend in clean energy finance is community co-ownership. Instead of a distant utility holding all the shares, local cooperatives or municipalities buy stakes, turning monthly bills into long-term investments. In Ontario, several towns investing in the Darlington project will share in its revenue, keeping profits — and decisions — close to home.

For families, this transforms a power plant from an external supplier into a collective asset. Parents can point to the steady light in their child's bedroom and say, "We helped build that." Pride and accountability grow together.

Education and Careers for the Next Generation

The spread of SMRs is already reshaping education. Technical colleges offer new programs in advanced welding, modular construction, and reactor operations. High schools create "energy academies," blending science with internships. In Poland, vocational students ride buses to a construction site each week, returning with stories of robotics and control rooms. In South Africa, apprenticeships train electricians and plumbers to maintain neighbourhood-scale reactors.

Parents see these opportunities as more than careers. They are lifelines that let young people stay in their hometowns, carrying local identity into a cleaner future.

Culture, Art, and Everyday Identity

As SMRs take root, they leave marks beyond the grid. Murals of constellations and stylized atoms appear on cooling-tower walls. Towns host annual "Light Days," celebrating the date their reactor first delivered power with music, food, and science fairs. Children invent playful nicknames — tiny sun, pocket power, Lego reactor — that slip easily into bedtime stories.

These small cultural shifts matter. They weave the technology into daily life the way river dams once defined fishing towns or windmills defined coastal villages. Energy becomes not just something a family consumes but something it belongs to.

International Lessons and Shared Wisdom

Because SMRs are appearing in such diverse places — Arctic villages, European industrial hubs, Pacific islands — families are already trading advice across borders. School classes in Ontario video-chat with peers in Estonia and Kenya to compare safety drills and emergency plans. Parents in Alaska swap tips with families in northern Sweden about preparing for winter logistics.

This quiet network of shared wisdom is perhaps the most hopeful sign of all: *the energy transition as a global conversation among ordinary households*, not just experts and politicians.

A Clean Legacy

For all their technical sophistication, the most powerful promise of SMRs is emotional. They allow parents to tell their children a hopeful story: that we faced a climate crisis and found ways to keep the lights on without darkening the sky; that we built technology carefully, shared its benefits fairly, and left the world cleaner than we found it.

Families know that real change is measured not only in megawatts but in daily life — in the air a child breathes, the pride of steady work, the trust between neighbours. Small Modular Reactors will never be the only solution to the energy challenge, but they are quickly becoming one of the most human.

In the decades ahead, these compact powerhouses may prove that "nuclear" need not mean massive or distant. Instead, it can mean local, inclusive, and quietly transformative — an energy future built not only for families, but with them.

Family Action:

If an SMR is proposed near you, ask about safety design, waste handling, local jobs and training, and how community concerns will be addressed.

CHAPTER 24

Working Together with Renewables

How nuclear and renewables can complement each other

Too Often, a False Choice

When families first begin talking about clean energy, the conversation often feels like a competition. Should we cover every roof with solar panels? Should wind turbines shoulder the entire load? Should nuclear keep its place on the grid or fade away?

This "either-or" framing makes for lively debates, but it misses the heart of the matter. The real question is not which technology wins. It is *how the clean technologies we already have can work together to give families both breathable air and electricity that never wavers.*

In chapter 9, we sampled the full energy menu. Now we focus on two of the most important items on that table — *nuclear power and modern renewables* — and show why they are partners rather than rivals.

Why Partnership Matters

Every household understands the strength of partnership. Parents divide chores so dinner arrives on time. Children team up on a school project. Neighbours lend each other generators when a storm takes down power lines.

Energy works the same way. Solar panels and wind farms shine when nature cooperates, flooding the grid with clean electrons. But the sun always sets and the wind sometimes falls silent. When that happens, families still need hot meals, warm bedrooms, and working hospitals. *Nuclear power provides the steady heartbeat that keeps those essentials running through calm nights and winter storms.*

Together, renewables and nuclear create a grid that is both clean and resilient. Renewables offer flexibility and rapid growth. Nuclear supplies a constant backbone. Families don't need to pick a favourite; they need the *teamwork of both.*

Ontario: Breathing Easier After Coal

A generation ago, families in Ontario faced smog days that turned summer skies brown. Children with asthma carried inhalers to playgrounds. Parents checked air-quality alerts before opening windows. Coal plants kept the lights on, but at a heavy cost.

The province decided on a daring plan: *Phase out coal entirely.* Wind turbines rose across rural ridges, solar panels appeared on schools, and nuclear plants quietly provided round-the-clock baseload electricity. By 2014, every coal unit was closed.

The results were immediate and personal. *Smog days disappeared.* Children could sprint through sprinklers without coughing. Hospitals recorded fewer respiratory emergencies. Families felt the difference not as an abstract climate metric but as clear lungs and bright skies. Ontario's story shows that *when nuclear and renewables pull together, they can do what neither could achieve alone — eliminate coal without sacrificing reliability.*

Denmark and France: Different Paths, Shared Lesson

Denmark's commitment to wind has made it a national emblem, visible across skylines and shorelines alike. On strong wind days, electricity flows outward to neighbouring countries. But when the air stills at night, Denmark relies on imports to keep the system balanced — a contrast that becomes especially clear when set beside France's steady nuclear supply. The comparison highlights not competition, but complementarity.

France, by contrast, is a land of reactors. Since the 1970s oil crisis, *nuclear stations have supplied roughly 70 percent of its electricity*, giving families low-carbon power and stable bills. Still, when winter demand surges, France sometimes buys wind-rich electricity from Denmark and its neighbours.

The quiet truth is that *these two countries depend on each other.* Denmark's breezy days complement France's steady nuclear output; France's reactors backstop Denmark when the air is still. The partnership crosses borders, proving that *no single source — however mighty — can stand entirely alone.*

California: Learning the Hard Way

California is famous for sunshine, and families there have embraced it with enthusiasm. Rooftop panels glitter on suburban streets, and vast solar farms stretch across desert basins. On bright afternoons, the state can generate more solar electricity than it can use.

But when twilight falls and dinner ovens switch on, the solar supply plunges just as demand peaks. Blackouts and price spikes have followed. Years ago, California planned to close Diablo Canyon, its last nuclear plant. Families debated: Was it worth keeping?

As outages grew and climate goals sharpened, the answer changed. *Diablo Canyon won a new lease on life*, not to compete with solar but to complement it — *a steady partner for the hours when sunshine fades.* Parents explain it simply to their kids: "The solar panels power our house all day. Diablo Canyon keeps the fridge running at night."

Other Places, Same Pattern

The partnership of renewables and nuclear is not unique to Ontario, Denmark, or California. Around the world, families are discovering that blending these technologies creates the most reliable path to a clean future, even though the mix of ingredients varies from place to place.

Consider South Australia, where rooftop solar has become almost a household signature. On bright afternoons, panels on homes and schools generate more than half of the state's electricity. Parents can glance at a phone app and see that their own roof is helping to power the grid. Yet when evening comes and the sun disappears, that same grid leans on neighbouring regions for steady supply. Hydropower from Tasmania flows across an undersea cable, and nuclear from other Australian allies fills in the gaps. Families there enjoy the confidence that their lights will stay on long after the last rays fade, proof that solar abundance and imported steady power can live side by side.

Across the Pacific in Japan, a different balance is taking shape. After years of debate following the Fukushima disaster, the country is cautiously restarting some of its nuclear reactors while also racing to build offshore wind farms along its long coastline. Parents who once feared radiation now talk with their children about the need to reduce costly fossil fuel imports and cut carbon pollution. Offshore turbines promise clean electricity when coastal winds are strong, while reactors supply dependable power through calm spells and typhoon seasons alike. Japanese families, who know the price of both blackouts and climate change, find reassurance in this careful blend.

South Korea offers yet another variation on the theme. There, dense cities and a powerful manufacturing sector demand electricity around the clock. The government is pushing ambitious solar targets — panels now gleam from apartment towers and factory rooftops — while also maintaining one of the world's most advanced nuclear fleets. Families can see the results in their own neighbourhoods: solar arrays soaking up daylight, reactors quietly providing heat and light after sunset. Parents tell their children that these twin sources keep their schools warm in winter, their subways running on schedule, and their air far cleaner than it was a generation ago.

The details differ — different climates, different politics, different histories — but the lived experience converges. Families from Adelaide to Osaka to Seoul breathe cleaner air, pay steadier bills, and sleep more easily knowing that when the wind drops or the clouds roll in, the grid will not fail them. Whether the backdrop is southern hemisphere sunshine, Pacific trade winds, or bustling industrial corridors, the pattern remains the same: Renewables bring the sparkle of innovation, nuclear provides the backbone of reliability, and together they give households the simple gift of dependable power without the shadow of smog.

Why Families Care About Resilience

When a winter storm howls or a heatwave strains air conditioner, families care less about which technology sends electrons through the wires and more about *whether the lights stay on.* Renewables excel when nature cooperates; nuclear excels regardless of weather. Together they offer the calm parents crave when refrigerators must stay cold and medical devices must keep running.

Resilience is more than engineering. It is emotional security. Knowing that a grid has both flexibility and backbone lets families plan meals, charge phones, and sleep through the night without anxiety.

The Moral Case for Teamwork

For decades, activists and politicians cast nuclear and renewables as enemies, urging citizens to "pick a side." But climate change is an unforgiving opponent. Air pollution kills millions each year. *Families cannot afford an energy civil war.*

The ethical path is cooperation. Renewables deliver growth and local participation. Nuclear offers scale and unwavering reliability. Together they form a system that is not only low-carbon but fair — one that respects the right of every child to breathe clean air and every parent to count on steady power.

Parents often explain it to their children in the simplest way:

"The sun gives us power when it shines.
The wind gives us power when it blows.
Nuclear gives us power all the time.
We need all three to keep the world bright and safe."

From Rivalry to Partnership

Chapter 9 highlighted that every energy source comes with its own strengths and limitations. Chapter 24 adds a deeper insight: Nuclear power and renewable energy reach their fullest potential when they work together. Ontario demonstrated this by eliminating coal. Denmark and France show it through cross-border electricity trading. And California proves it every evening, as steady nuclear output carries the grid after the sun sets.

The future will not be written by rivalry. It will be written by *partnership* — a partnership strong enough to light homes, clear skies, and give families everywhere the confidence that tomorrow's power will be both clean and sure.

Families need not choose between the sun and the atom. They can, and must, have both — because the clean, reliable world our children deserve will be built by every source of wisdom and every watt of cooperation we can muster.

Family Action:

- Learn how different clean energy sources support one another in your region.
- Visit a local wind farm, solar site, or nuclear facility to see partnership in action.
- Share what you learn with neighbours and community groups, and start family conversations that move beyond "either-or" thinking toward "How can we work together for reliable, clean power?"

Part V

Energy, Culture, and Imagination

This section shifts from the *technical and policy* lens into the *cultural and human* one. Energy here is about identity, vision, and imagination: how families weigh values, how the arts shape feelings, how culture on-screen moulds perceptions, how innovators inspire, and how we look ahead together.

CHAPTER 25

A Family Decision for Our Future

Balancing risks, rewards, and values

Conversations at the Kitchen Table

Most energy debates begin not in parliaments or corporate boardrooms, but in kitchens and living rooms. Families gather around tables scattered with bills, homework sheets, and glowing phones, trying to decide what kind of world they want to hand to their children. Energy is vast and global, yet also intimate — an invisible current running through bedtime stories, morning coffee, and quiet weekend meals.

In Toronto, the Singh family lingers after dinner. Their teenage son, Aarav, has just read an article about nuclear power and asks, "Why do we still use it if people say it's dangerous?" His father gestures toward the clear sky outside, remembering the smog days that disappeared when coal plants shut down.

"This," he says, "is what nuclear gave us — clean air."

His mother adds gently, "It isn't only about safety. It's about the kind of future we want you to inherit."

Later, in the Schneiders' household, the conversation returns — this time with sharper edges. Lena asks why reactors were shut down if coal plants eventually took their place. Her parents speak of Chernobyl and Fukushima, of protests and promises, of a nation that believed rejecting nuclear was an act of care. "It felt like protecting you," her mother says quietly. Lena looks down at her plate, then back up. "But if coal came back," she asks, "was that really safer for me?" The question hangs in the air, unanswered, heavier than the plates between them.

In South Africa, the Mokoena family eats by candlelight during another bout of load-shedding. Their teenage son sighs, "Why can't we have power like other countries?"

His father shrugs. "Because our leaders failed us. But maybe nuclear could help."

His wife counters, "Yes, but at what cost — and will it reach us or only the wealthy?"

Their words capture a nation's dilemma: hope for reliable power tempered by memories of broken promises.

From Berlin to Nairobi to Buenos Aires, families share the same rhythm of questioning. What risks will we accept? What rewards do we seek? Which values are non-negotiable? Energy is not merely a technical debate. *It is a family decision.*

Balancing Risks and Rewards

Every parent understands that life is a balance of risk and reward. We send children to school, trusting buses and teachers. We drive cars, fastening seat belts and hoping other drivers stay alert. We vaccinate, weighing tiny risks against the immense protection they bring. Energy choices are no different.

Nuclear power makes that calculation especially vivid. Its risks — accidents, long-lived waste — are dramatic and easy to picture. Its rewards — clean air, reliable baseload electricity, climate protection — are experienced quietly every day.

In Ontario, parents remember what came after the coal phase-out: children breathing easier, skies turning a pure blue. In France, the reward is stability — predictable bills and independence from volatile gas markets. In Fukushima, by contrast, the trauma of evacuation remains personal. Families fled homes and schools with little warning. Even though radiation exposures proved lower than feared, the fear itself lingers.

Risk and reward shift depending on the comparison. In Poland, where coal still dominates, many parents see nuclear's dangers as manageable next to the daily certainty of soot and asthma. In Germany, with its strong renewable sector, nuclear feels less essential and its risks loom larger. In South Africa, where rolling blackouts darken kitchens night after night, the promise of nuclear reliability outweighs nearly every other concern.

Families weigh these factors instinctively, guided by values: Protect the vulnerable, honour ancestors, give children the best chance at a stable life.

Values We Carry Forward

When families decide which energy to support, they are really deciding which values will shape tomorrow.

For Indigenous peoples, energy is inseparable from relationship with the land. In Cree communities of Canada, elders tell children that rivers are not resources but relatives. Dams that flood valleys are not just technical projects but cultural wounds. In New Zealand, Māori families view geothermal springs as *wāhi tapu* — sacred places. Harnessing them requires ceremony and guardianship, teaching children that tradition and technology can coexist.

Faith traditions reinforce similar lessons. Catholic families recall Pope Francis's "Laudato Si'," which calls for stewardship of creation; wasting electricity becomes more than thrift — it becomes a spiritual duty. Muslim households speak of *khalifa*, humanity's role as guardian of the Earth. Hindu families along the Ganges regard polluting the river as a moral failing, not merely an environmental one.

Secular families too root energy choices in ethics. Many focus on fairness across generations. Parents plant trees whose shade they will never enjoy, or invest in home insulation that will mainly benefit their grandchildren.

"Someone planted trees for us," a German father tells his son. "Now it's our turn."

In every culture, energy decisions are cultural decisions — about reciprocity, respect, and justice as much as megawatts.

Children as Catalysts

Children are often the boldest voices at the table. They ask the questions adults sometimes avoid.

After climate strikes swept through Europe, parents in Hamburg and Paris heard their children demand, "Why are we still burning coal if it ruins our future?" In Kenya, students who joined the "solar scholars" program returned home eager to wire panels for their families. "She became our teacher," one father said proudly.

In Berlin, a teenager confronted her parents: "If you knew something was hurting my future, wouldn't you stop it?" The question reframed the entire conversation.

Children's urgency cuts through statistics and politics, reminding parents that energy choices are acts of love — or neglect — toward the next generation.

Stories of Choice Around the World

Every nation's path to cleaner energy is ultimately written in the daily decisions of its families, and each place carries its own history, hopes, and anxieties.

In Ontario, the story is one of sacrifice that blossomed into relief. When the province committed to shutting down coal plants, many parents braced for higher short-term costs and worried about grid stability. Some families even installed backup generators, fearing brownouts. But when the last coal plant closed and the nuclear fleet quietly shouldered most of the load, a new kind of conversation began around dinner tables. Skies that had once turned a murky grey in summer cleared to a bright blue. Asthma attacks dropped. Parents found themselves telling children,

"We paid a little more and planned a little harder, but we gave you air you can run through without coughing."

Across the Atlantic, France tells a different energy story. In towns like Chinon or Gravelines, cooling towers rise alongside church spires and harbour cranes, woven into the familiar landscape. Older generations still speak of the Messmer Plan as a turning point, recalling a moment when the country chose to rely on itself. When energy prices swing wildly elsewhere, French families often feel the steadiness of that earlier decision in small but meaningful ways — predictable bills, reliable power, and the quiet sense that a choice made decades ago continues to shelter the present.

Cross into Germany, and the mood changes. The decision to shut down reactors after Fukushima is remembered not merely as policy but as a moral act. Families who marched in the anti-nuclear protests of the 1980s tell those stories with the same gravity others reserve for civil rights marches. Parents explain to their children that refusing nuclear power was a way of saying, "We will not gamble with our grandchildren's safety." Yet the return of coal and the pinch of expensive imported gas complicate the narrative. Around many tables, a subtle debate unfolds between generations: Older parents defend the phase-out as a matter of principle, while teenagers, anxious about climate change, ask whether the principle still makes sense if it means burning more fossil fuel.

In Japan, memory itself is the central character. The trauma of March 11, 2011 — earthquake, tsunami, and the cascading failure at Fukushima — still hovers in family conversations. Parents recall hurried evacuations, children clutching a single toy as they left their homes. Even families far from the plant remember the uncertainty, the empty store shelves, the mistrust. A decade later, some households argue that nuclear power must never return; others, staring at high energy prices and a warming planet, quietly wonder if carefully regulated reactors are part of their children's best chance. Dinner-table debates often end with long silences, as memories of fear wrestle with the pressure of necessity.

Far to the south, in South Africa, the story is shaped by darkness — literal darkness when rolling blackouts, known as load-shedding, sweep through neighbourhoods night after night. Children do homework by candlelight, freezers defrost, and small shops lose their day's earnings. Families speak of frustration and betrayal, of government promises unkept. Yet they also speak of hope. For some, nuclear energy represents a lifeline: a technology that could finally keep the lights on. Others worry about cost, corruption, and whether the benefits will reach the poorest households. Their conversations capture a nation balancing the hunger for reliable power with the scars of political mistrust.

And in the scattered Pacific Islands, families tell stories of land itself slipping away. Grandparents in Kiribati or the Marshall Islands describe beaches that have already vanished, coconut trees now half-drowned by rising seas. For them, the climate crisis is not a projection; it is a tide lapping at the doorstep. Clean energy is framed not as a policy choice but as an act of survival. At the same time, older islanders remember the era of nuclear testing, when distant powers used their atolls as laboratories and left radiation in the soil. That memory complicates every discussion of modern nuclear power: Some elders reject it outright, while younger voices — desperate to slow the rising ocean — ask whether new, safer reactors might be different. Around their tables, survival and skepticism wrestle like ocean currents.

These narratives differ in detail but echo one another in spirit. Whether it is a Toronto parent celebrating clear skies, a French family proud of national independence, a German teenager questioning an inherited moral stance, a Japanese household haunted by disaster, a South African family yearning for steady electricity, or Pacific Islanders fighting for the very ground beneath their homes, the conversation always circles back to the same quiet question: *What do we owe our children, and how brave are we willing to be in delivering it?*

Energy as a Family Decision

Governments may set policies and corporations may build plants, but families make countless choices that shape the energy landscape. They decide whether to install rooftop solar, buy an electric car, or vote for leaders who champion specific technologies. They show up at town halls, demanding fairness or voicing support.

Even the smallest acts — switching off lights, explaining a utility bill to a curious child — carry meaning. Each is a lesson in responsibility, a tiny declaration of values.

Energy is a family decision, not because each household controls the grid, but because *each household decides the principles that guide the future.*

The Legacy We Leave

At the heart of every conversation lies a simple question: *What legacy will we leave our children?* Families may disagree about nuclear or renewables, about costs or timetables, but they share a desire to protect the next generation.

A family decision for the future means accepting that no choice is perfect, but inaction is its own peril. It means balancing today's fears with tomorrow's hopes. It means telling children not only what technologies we chose, but *why* we chose them — fairness, stewardship, justice, survival.

Energy will always involve engineering and economics. But for families, it will always be personal and moral. The world our children inherit will be written not only in government plans or corporate strategies, but in these everyday kitchen-table debates where parents and children weigh risks, rewards, and values — and quietly decide the shape of history.

Families are not merely consumers of electricity. They are the authors of the story of energy — choosing, every day, what kind of sky their children will breathe beneath, and what kind of future will follow.

Family Action:

Gather your family to weigh the risks and rewards of how you power your life. Talk about what values — safety, fairness, or stewardship — matter most, and let those guide how you choose, vote, and plan for the energy future you want to leave behind.

CHAPTER 26

Energy and the Arts — Feeling the Future

Songs, stories, and images of power

Why the Arts Matter in Energy

Facts can inform, but art moves hearts. Families may learn statistics about emissions or efficiency, yet what lingers are the songs sung at protests, the novels read aloud in living rooms, the murals glimpsed on the walk to school. Art translates energy into feeling, turning invisible kilowatts into something a child can imagine and a parent can cherish.

When a mother hums a ballad about coal miners, or when a child reads a science fiction story about a glowing city powered by fusion, energy becomes emotional. It enters memory and family lore. Engineers may design tomorrow's power plants, but artists give families the pictures and melodies that make those plans meaningful. A single line of poetry or a haunting melody can hold the urgency of climate change more powerfully than pages of charts.

Art also provides a bridge across generations. Grandparents who once sang mining songs can find common ground with grandchildren who sketch wind turbines in art class. Parents who grew up with 1980s anti-nuclear posters now watch their own children create animations about solar-powered cities. In this way, art carries both history and possibility, ensuring that the emotional lessons of the past travel forward even as technology changes.

Songs of Power and Protest

Music has always been tied to the work of energy. Coal-mining ballads once echoed through Appalachian kitchens, sung by parents who faced darkness underground and uncertainty above. Songs like "Dark as a Dungeon" captured both the dignity and the danger of digging fuel from the earth — stories children absorbed even if they never set foot in a mine.

Across the Atlantic, Welsh choirs filled chapels with hymns about miners' pride and peril. In South Africa, protest songs during apartheid tied electricity to freedom. Families sang to demand fairness, light, and justice; the act of singing itself became power, a reminder that even without light, voices could shine.

Today, climate movements carry their own soundtrack. Youth-led strikes ring with chants and harmonies. Parents hear them drifting through city squares or on social media; children hum them on the way home from school. Kenyan students write rap verses about solar lamps that light evening study sessions. Icelandic folk singers weave wind turbines into haunting ballads. Rock bands stage concerts powered entirely by solar arrays, while choirs perform under wind-lit stages. Around the world, music gives energy a rhythm families can march to, dance to, and remember, turning policy debates into shared emotional experiences.

Stories in Literature

Energy has always fueled stories. In nineteenth-century England, coal powered Dickensian London, where novels such as *Hard Times* described smoke-darkened skies and families squeezed by factory schedules. Readers felt the tension of progress and pollution long before environmental science had a name.

The atomic age brought a new kind of narrative. Mid-century science fiction imagined glowing cities with limitless clean power — and dystopias ravaged by radiation. From *The Day of the Triffids* to Japanese manga about post-nuclear worlds, families debated at the dinner table whether nuclear was salvation or threat.

Today, children's literature introduces renewables as characters in their own right. Picture books paint wind turbines as friendly giants and the sun as a benevolent provider. In a popular Kenyan storybook, a boy builds a solar lamp for his mother so she can weave baskets at night; real children copy the project, turning fiction into action. Young adult novels now feature heroes who repair microgrids or pilot solar airships, making clean energy part of everyday adventure.

Families reading these stories together pass on more than entertainment. They hand down metaphors and mental pictures that guide future choices. When energy is framed as monster or hero, threat or promise, those frames linger long after the final page.

Murals, Paintings, and Images

Art also lives on walls and canvases, silently shaping public feeling. Murals in former coal towns often depict miners with lamps and tools — honouring sacrifice while hinting at the cost of the sacrifice. In renewable-rich regions, families see bright images of sunbursts and whirling turbines.

In Mexico City, a huge mural shows school child holding panels that bloom like flowers. In Morocco's desert towns near the Noor solar complex, artists paint giant suns that spill over building façades, celebrating a new economic lifeline. Toronto boasts a community mural where the blue of Lake Ontario blends with the silver domes of nuclear plants, surrounded by fish and families. It reframes reactors not as threats but as neighbours keeping the air clear. In Germany, anti-nuclear art from the 1980s still shines on village walls: yellow suns bearing the phrase "*Atomkraft? Nein danke!*" (Nuclear power? No thanks!). Children who see these images ask questions their grandparents once asked, proving that murals can outlast speeches and even technologies.

A single painting of smokestacks or a bright poster of solar rooftops can shift a child's feelings for years, creating an emotional memory no statistic can rival.

Cinema and Theatre as Cultural Mirrors

Film and theatre turn energy into shared cultural experience. Postwar Japanese movies, haunted by Hiroshima and Nagasaki, framed radiation as terror, embodied in monsters like Godzilla. For families, these films made nuclear power synonymous with destruction.

Modern filmmakers imagine different futures. Animated stories such as *WALL-E*, *Moana*, and *Howl's Moving Castle* frame energy as a moral choice: Waste leads to collapse, stewardship to balance. Families watch these films together and absorb their quiet lessons about care for the planet.

Theatre has played its part too. British plays like *The Pitmen Painters* dramatize miners' lives and the dignity of work. In Canada, productions about hydroelectric projects explore the tension between development and Indigenous rights. School drama clubs from Chicago to Nairobi now stage original pieces about climate change and solar dreams, giving young actors — and their parents in the audience — a direct emotional stake in energy's future.

Children's Drawings and Imagination

Perhaps the most revealing art about energy is made by children themselves. Ask a classroom to draw "the future of power," and many will sketch solar panels on rooftops, wind turbines on hills, or glowing stars representing fusion. Few draw coal. Some show nuclear plants as ominous grey towers; others as sleek blue domes beneath cheerful skies.

Teachers notice that children often personify energy: the smiling sun, the playful wind, the steady river. Families who hang these drawings on refrigerators see how deeply kids connect energy to feeling. In one Ontario school, a child drew a family plugging an electric car into a giant daisy — an image both whimsical and prophetic. In rural Kenya, students decorate exercise books with bright yellow panels and tiny windmills, showing how new technologies quickly enter daily imagination.

These drawings matter because they reveal a cultural shift already underway. The future is not only engineered in laboratories; it is being coloured with crayons in classrooms and taped proudly to kitchen walls. When children imagine clean skies and friendly turbines, they normalize those visions for their parents as well. Family conversations change, and so do expectations of what is possible.

Energy as Metaphor in Poetry and Faith

Poets have long turned to energy as metaphor — the fire of passion, the spark of creativity, the fuel of love. Those images slip easily into family language: "You light up my life," "She's a live wire," "He burned out." Such phrases keep energy close to the heart.

Faith traditions do the same. The Christian Bible speaks of light as truth and fire as spirit. Islamic teachings describe humanity as *khalifa*, guardians of the Earth's balance. Hindu stories celebrate the sun as the eternal giver of life, while Buddhist texts use the turning wheel — an ancient turbine — as a symbol of renewal. Indigenous prayers give thanks for wind, fire, and water. Parents who pass these stories to children are teaching energy literacy as much as spiritual wisdom, framing stewardship not as duty imposed but as love practiced.

Across the Pacific, Hawaiian chants honour the trade winds that carry both canoes and electricity through modern turbines. In the Andes, Quechua festivals celebrate Inti Raymi, the Sun Festival, where dancers weave ancient reverence into modern calls for solar power. These rituals teach children that caring for energy sources is a sacred trust, blending science and spirituality.

Global Voices of Art and Energy

Across the world, families experience energy through art that reflects local struggles and hopes. In Norway, Sami painters place reindeer against backdrops of wind turbines, questioning what "green" progress means for ancient herding routes. In Kenya, folk songs celebrate the solar lamps that replaced smoky kerosene, their upbeat rhythms turning new technology into community pride. In the Marshall Islands, poets describe rising seas and nuclear testing scars, linking climate, survival, and memory.

In the Middle East, contemporary calligraphers weave solar motifs into traditional Arabic scripts to symbolize renewal. Brazilian street artists paint hydroelectric dams as both threat and promise, capturing the double edge of progress. In India, Bollywood filmmakers have begun to feature solar rooftops and electric buses in everyday street scenes, quietly shifting what millions of viewers see as modern and desirable. In Canada, Indigenous artists carve wind and wave patterns into cedar panels, pairing ancestral respect for nature with visions of renewable futures.

These works carry energy debates far beyond expert reports. Parents and children engage not through policy papers, but through songs, stories, films, and images that stir emotion and invite conversation around the dinner table. Art turns distant technologies into intimate choices.

Why Families Need Feeling as Well as Facts

Families need both information and imagination. Facts explain how grids function; feelings explain why choices matter. A father might be convinced by statistics that nuclear power is safe, but a protest song from his youth still makes him cautious. A child may not grasp carbon math, but a picture book about rooftop solar can spark a lifetime of climate action.

Art does not replace science, but it gives science resonance. It helps families act not only on logic but on culture, weaving knowledge into daily life. A parent who remembers a childhood mural of clean rivers is more likely to support water-protecting policies. A teenager inspired by a graphic novel about fusion-powered travel may choose a career in physics. These emotional connections turn abstract numbers into lived motivation, ensuring that technological progress is accompanied by shared meaning.

Feeling the Future

Energy isn't just about wires, bills, or emissions. It is about how we imagine the world we want. Art helps families feel that imagination. It frames nuclear as pride or peril, renewables as hope or struggle, climate action as duty or dream.

When families teach children energy facts, they give them knowledge. When they share energy art, they give them vision. Both are needed. The future will not be built only by engineers and policymakers. It will also be imagined by artists — and by families who carry those songs, stories, and images into the generations to come, singing and sketching the energy future long before it is fully built.

Through art, a parent can comfort a child anxious about climate change or inspire a teenager wondering what to study. Through music and stories, a family can see itself as part of a shared global effort rather than as an isolated household. Feeling the future, together, is the first step to creating it.

Family Action:

Create an at-home energy art project by repurposing materials to depict your family's clean energy future and sharing the story behind it.

CHAPTER 27

Energy and Culture On-Screen

Films, games, and popular imagination

Why Screens Shape Our Energy Imagination

Evenings in most homes begin with a glow. Families gather around televisions in living rooms, laptops on desks, or smartphones balanced on pillows. Through these screens, stories about energy unfold with a power that often eclipses any textbook or policy paper.

A parent may read careful statistics about nuclear safety, but the haunting images of the HBO miniseries *Chernobyl* stay longer in memory. A child may not understand how a photovoltaic cell works, yet after watching Moana restore harmony to her island, clean power feels magical and morally right. A teenager might never open an energy-policy report, but after hundreds of hours in post-apocalyptic video games, they instinctively link "nuclear" with collapse.

Screens bypass statistics and move straight to emotion. They are the modern campfire where today's folklore is forged, shaping how families picture the risks and promises of energy long before a single real-world decision is made. For older generations, early encounters with energy often came through screens — grainy footage of atomic tests, steam engines, or industrial power. Those first images linger, shaping how grandparents explain energy to grandchildren and which futures feel reassuring or unsettling.

Nuclear On-Screen: Fear and Fascination

Since the mid-twentieth century, nuclear energy has been one of cinema's most enduring symbols — sometimes of promise, but more often of peril.

In post-war Japan, *Godzilla* emerged from the sea as a monster awakened by radiation. Parents and children watching those black-and-white films did not need a lecture to understand the metaphor. Nuclear power meant destruction, an echo of Hiroshima and Nagasaki. Decades later, animated reboots keep the allegory alive, ensuring that even digital-native children absorb the cautionary link between radiation and danger.

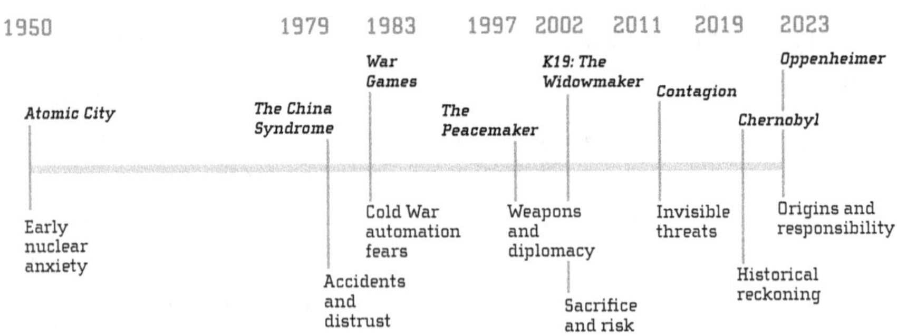

Fig. 20. Nuclear energy on screen: Fear and fascination over time.

American films of the Cold War reinforced that unease. Popular entertainment leaned heavily into nuclear extremes. Thrillers and dramas returned again and again to visions of catastrophic release and irradiated landscapes, reinforcing nuclear power as spectacle — something awe-inspiring, dangerous, and rarely ordinary. *Dr. Strangelove* turned nuclear brinkmanship into absurd dark humour, while *The Simpsons* introduced a generation to a bumbling safety inspector named Homer, asleep at the control panel. Reactors and weapons blurred together in the public imagination.

Yet fascination persisted alongside fear. Educational reels in the 1950s showed neat diagrams of atomic power lighting homes. When presenters or films linger on the contrast — a speck of fuel set against mountains of coal — children react instinctively. The comparison alone is enough to spark awe, curiosity, and sometimes unease, reinforcing nuclear's place in the imagination as something both powerful and hard to grasp. More recent documentaries reveal quiet awe as cameras linger on the elegance of containment domes and the precision of control rooms. When *Chernobyl* aired in 2019, it did more than frighten; it reopened conversations about trust, transparency, and the difference between technology and human failure. Families who had never thought about radiation suddenly debated evacuation plans at the dinner table.

Across continents, parents notice a dual response in children: a shiver at the idea of invisible danger and a spark of curiosity about such concentrated power. Nuclear on-screen remains a paradox — at once a marvel of human ingenuity and a reminder of human fallibility.

Renewables On-Screen: Hope and Harmony

Renewables, by contrast, are usually portrayed as natural, harmonious, and almost always benevolent.

In children's programs, wind turbines spin like friendly giants on sunny hillsides. Animated features imagine glittering solar cities where rooftops gleam like jewels. Documentaries such as *Before the Flood* and *An Inconvenient Truth* pair images of spinning blades and shining panels with uplifting music, offering viewers not just information but a sense of rescue.

Sometimes the symbolism is subtle. In *Moana*, the ocean itself becomes a character, teaching respect for natural balance. In *WALL-E*, the absence of balance — mountains of trash, skies choked with haze — drives the plot and the moral. Children leave the theatre asking whether our own planet could share that fate and what role wind, sun, and stewardship might play in preventing it.

Streaming services now carry documentaries about African villages lit for the first time by community solar, or about Arctic communities blending wind and micro-hydro. Families in distant cities watch together and imagine their own rooftops shimmering with panels. Across cultures, renewables arrive on-screen cloaked in virtue: They promise abundance without guilt.

Fossil Fuels On-Screen: Progress and Pollution

Fossil fuels tell a more complicated story.

In early cinema, coal and oil were heroic. Trains racing across wide landscapes, derricks gushing black gold, factories pounding with steel — all signalled progress and prosperity. Parents once pointed to these scenes with pride: Energy meant jobs and a modern life.

As awareness of climate change grew, the imagery darkened. Documentaries now linger on smokestacks under blood-red sunsets, pipelines snaking across fragile tundra, oil spills spreading like bruises on the sea. Films such as *There Will Be Blood* portray oil as both triumph and tragedy, progress entangled with corruption. Families watching absorb the ambiguity: Fossil fuels built the modern world but carry a heavy, lasting cost.

Even children's films echo the shift. In *Cars 2*, the villain is an oil baron disguised as a green entrepreneur. Young viewers laugh at the chase scenes but remember the message: Unchecked fossil fuel dependence is risky and dishonest.

Games: Interactive Energy Stories

Video games add a unique dimension because players do more than watch — they participate.

The *Fallout* series immerses teens and adults alike in a landscape devastated by nuclear war. Hours of scavenging through radioactive ruins can leave players with a visceral association between "nuclear" and catastrophe. Parents sometimes find themselves fielding tough questions about whether real-world reactors could ever cause such devastation.

Other games become quiet classrooms of energy literacy. In *SimCity* or *Cities: Skylines*, players must decide whether to build coal plants, nuclear stations, or solar farms. Each option carries different costs, risks, and reliability. Children argue strategy across kitchen tables: "Nuclear gives more power but might melt down. Wind is clean but needs backup."

Even the sandbox world of *Minecraft* hosts elaborate mods where kids craft solar arrays or carefully manage nuclear reactors. A ten-year-old might look up from the screen and ask, "Why don't we use reactors like in my game? They give infinite power." Games like these turn abstract trade-offs into personal experiments and bring those lessons to family conversations.

Virtual reality experiences are adding new layers. Museums now offer VR tours inside fusion experiments or offshore wind farms, letting families "walk" through energy futures that feel tangible and real.

Television: Energy in Everyday Life

Television continues to shape perception through both entertainment and news. For decades, *The Simpsons* kept nuclear safety a running joke. News broadcasts, drawn to disaster, highlight accidents far more than quiet success, reinforcing public anxiety.

Yet TV also educates. Public broadcasters air documentaries about solar microgrids in Africa or community wind projects in Europe. Families who watch together often pause to talk: *Could we do something like that here?* In some countries, reality programs even follow families installing rooftop panels or retrofitting old homes, turning efficiency into relatable drama.

Internationally, energy-themed dramas have become surprise hits. A Korean series about a family displaced by a reactor accident sparked debates across Asia about energy security. A French miniseries following a team of young engineers designing next-generation reactors turned into a youth recruitment tool. Families not only watch but sometimes act — attending community meetings, writing letters, or simply rethinking how they use electricity.

Family Conversations Sparked by Screens

What matters most is not the screen itself but the conversations it sparks afterward.

After watching *Chernobyl*, families in France and Germany debated whether their own nuclear fleets were truly safe. Parents explained why their countries chose different paths; children asked what would happen if an accident occurred today.

After a family movie night with *WALL-E*, children in Toronto began collecting recyclables the next morning, worried about mountains of trash. A boy in Kenya, fresh from an evening of *SimCity*, asked his father whether their village could build wind turbines "like in the game."

In rural India, a family watched a regional-language documentary on solar microgrids and decided to pool savings with neighbours for a shared installation. In Brazil, a teen who followed an online series about Amazon deforestation convinced her parents to switch to a renewable-only electricity provider.

Screens shape imagination, but it is around the dinner table, in bedtime chats, and during car rides that those images are tested, questioned, and woven into family values.

Global Differences in Energy Narratives

Culture colours these screen stories.

In Japan, where memories of Hiroshima and Fukushima run deep, nuclear cinema often carries the weight of trauma and caution. In the United States, narratives are polarized: Nuclear is villain in apocalyptic fiction yet saviour in documentaries about climate solutions. In France, where reactors quietly supply most electricity, nuclear rarely features at all, reflecting normalization rather than fear.

Across much of Africa, films and television celebrate solar power as empowerment — stories of children studying by clean light after generations of smoky kerosene lamps. In the Middle East, satellite channels increasingly showcase vast desert solar farms as symbols of economic independence. Scandinavian dramas weave wind energy into backdrops of fjords and forests, while Indigenous filmmakers highlight the struggle to protect ancestral lands from both fossil and "green" megaprojects.

Latin American cinema offers its own lens. Mexican and Chilean directors blend magical realism with cautionary tales of mining and hydro dams, teaching young audiences that even "clean" energy can bring injustice if communities are ignored. In Australia, documentaries about reef bleaching and desert solar farms mix grief with gritty optimism, leaving families inspired but unsettled.

Families watching in each place absorb not just images but cultural values. Screen narratives reinforce national choices and, in turn, guide the next generation's expectations.

Screens as Energy Educators

Screens may not resemble classrooms, but they teach all the same. Films, shows, and games supply emotional shorthand: nuclear as peril or power, renewables as promise, fossil fuels as fading legacy. These impressions seep into voting booths, career choices, and everyday habits.

A father might be persuaded by safety data, yet a youthful memory of *Godzilla* keeps a flicker of unease alive. A teenager who spends nights in the glowing deserts of *Minecraft* sees solar panels not as policy but as possibility. Grandparents who once thrilled to early oil-boom westerns may feel conflicted as grandchildren cheer for wind farms on-screen.

Parents and children build their understanding together, blending facts with feelings. The next time a family discusses installing rooftop solar or supporting a local nuclear project, echoes of their shared screen stories inevitably shape the debate.

Imagining Tomorrow

The arts of screen culture remind us that energy is not only technical — it is imaginative. The stories families watch and play today become the mental blueprints for tomorrow's power systems.

When parents explain an electricity bill, they give knowledge. When children talk about *WALL-E*, debate reactor safety after *Chernobyl*, or design wind farms in a game, they give vision. Together, they are sketching the outlines of the future. And as with every great story, the ending is still being written — on our screens, in our homes, and in the choices each family makes.

Family Action:
Use films and games to spark discussion by asking what's realistic, what's exaggerated, and what the family learned about energy systems.

CHAPTER 28

Global Innovators and Voices

Profiles of energy heroes and changemakers

Why Innovators Matter for Families

When children hear about energy, they often picture giant machines — wind turbines taller than church steeples, nuclear domes gleaming beside rivers, endless solar farms shimmering in desert heat. Yet every one of those machines began as a human idea. Someone imagined it, someone argued for it, someone convinced a community to believe in it.

Families need these human stories. Innovators turn the abstract into the possible, proving that the energy transition is not only about technology but also about courage, persistence, and heart. Parents know that role models matter. Children admire athletes and musicians. Why not engineers, village leaders, and quiet neighbourhood heroes who are reshaping how the world is powered?

Energy pioneers remind families that the future is built not only in parliaments or corporate headquarters, but also in basements, classrooms, and small workshops — in places that look very much like home.

Youth Voices: From Ideas to Movements

Some of the most powerful sparks come from the youngest hands.

In southern India, fourteen-year-old Vinisha Umashankar watched coal smoke billow from roadside ironing carts and thought there must be a better way. Her solar-powered ironing cart — light, portable, and affordable — now inspires copycats across Asia. Families see her story and realize that innovation isn't reserved for adults in high-tech labs.

In Sweden, Greta Thunberg began her climate strike alone outside parliament. Within months, millions of students were skipping school to march. Parents who once thought climate change was distant suddenly faced children asking, "Why aren't you acting faster?"

Kenyan "solar scholars" bring panels to village schools, wiring lights so classmates can study at night. Parents call them "the children who brought light," proud that the next generation isn't waiting for government permission to create change.

Across the United States, high school robotics teams now build miniature wind turbines and compete nationally, proving that clean-tech excitement can rival any football game. In Brazil, teenagers design micro-hydro systems for rainforest villages, blending engineering with conservation.

From Jakarta to Johannesburg, youth clubs now pair climate action with music and theatre, using art to call parents and leaders to account. Youth voices bring urgency and imagination. They remind families that the transition is not about far-off generations — it is about the children laughing in our own kitchens today.

These global stories remind us that hope is already taking shape.

Inventors in Labs and Workshops

Innovation often begins in quiet corners: a basement, a garage, a small research lab lit late into the night.

In Canada, engineers refining Small Modular Reactors explain to their children that they are building not just a machine but a future where clean electricity can reach remote northern towns.

At California's National Ignition Facility, researchers recently achieved the long-sought "net energy gain" in a fusion experiment — more energy out than in. Families across the world read the headlines and asked at dinner, "Does this mean limitless clean energy?" Scientists replied: It's a first spark, but a spark that proves the dream is possible.

In Nigeria, entrepreneurs design pay-as-you-go rooftop solar kits so families no longer wait for distant utility companies. In rural Indonesia, women's cooperatives assemble and maintain small wind systems, earning income and respect. In Bangladesh, engineers created the world's largest solar home program, bringing light to more than four million off-grid households.

These inventors show that breakthroughs are rarely sudden miracles; they are years of trial and error, of parents balancing late-night experiments with bedtime stories.

Community Leaders and Cooperatives

Innovation is social as much as technical. Around the world, neighbours are rewriting energy by working together.

In many Danish towns, wind power is not only seen but owned. Local cooperatives allow residents to buy shares in turbines, turning neighbours into stakeholders. Children grow up watching their parents attend annual meetings that feel more like village celebrations than board sessions, reinforcing the idea that energy can be something communities build — and take pride in — together.

German villages blanket barns and schools with solar panels, selling power back to the grid. Residents speak of independence and collective responsibility; renewable power feels less like a corporate product and more like a community harvest.

In Canada, seventeen First Nations joined forces to create the Wataynikaneyap Transmission Project, bringing clean, reliable power to remote communities that once relied on costly diesel. Elders explain to children that this time energy came *with* them, not *to* them.

In the Philippines, island communities build microgrids powered by solar and storage so that typhoons no longer mean weeks of darkness. In Uruguay, entire neighbourhoods have pooled funds for shared solar arrays, meeting monthly to plan new expansions while children play under strings of efficient LED lights.

From cooperatives in rural Nebraska to fishing villages in Vietnam experimenting with tidal power, these stories teach families that power is not only generated in plants but also in people — when communities unite around a vision.

Voices of Justice and Equity

A clean grid means little if it is unjust. Innovators in equity are as vital as those in engineering.

In South Africa, civic activists fight to ensure that new wind and solar farms benefit township residents, not just wealthy suburbs. Parents insist their children deserve the same steady electricity as anyone in Cape Town's business district.

In Chile and Argentina, Mapuche leaders guard rivers and ancestral lands from poorly planned hydro projects, reminding the world that clean energy must still honour culture and consent.

Across the Pacific, poets and organizers in the Marshall Islands and Tuvalu speak of rising seas as both ecological and moral crisis. Their art and activism carry their tiny nations' voices to global climate talks, proving that even the smallest populations can shape the global agenda.

In the United States, environmental justice advocates in places like Louisiana's "Cancer Alley" demand that the new hydrogen and carbon capture industries protect low-income families from repeating the harms of the petrochemical era.

Justice innovators show families that energy is about dignity as much as supply.

Profiles of Energy Heroes

Across continents and cultures, certain individuals have stepped forward as vivid proof that ordinary people can redirect the world's energy story. Their journeys unfold in villages and cities, on windswept coasts and in quiet laboratories, and they remind families everywhere that innovation is as much about persistence and courage as it is about technology.

Consider James Hansen, the American scientist who first warned the United States Congress in 1988 that the planet was warming. He faced political headwinds and public doubt, yet he kept presenting data year after year. Parents who once dismissed climate change began to recognize his steady voice as a call to protect their children's future. His determination showed that science, when paired with moral clarity, can awaken an entire generation.

From the diplomatic halls of the United Nations came Christiana Figueres, the Costa Rican leader who guided nearly two hundred nations toward the Paris Agreement. She spent years coaxing rivals into cooperation, turning late-night negotiations into moments of fragile consensus. Families who followed her work on the news saw something rare: a global pact forged not by threats, but by optimism and relentless dialogue. Her achievement reassured parents that international cooperation, though difficult, can succeed when framed as a shared responsibility to the next generation.

In rural Malawi, a boy named William Kamkwamba refused to let poverty dictate his future. After dropping out of school, he scoured a library, studied diagrams, and scavenged scrap parts to build a working windmill that brought electricity to his village. Children across Africa now read his story in classrooms, seeing themselves in his resourcefulness. For many families, William's windmill is more than a machine; it is proof that imagination and determination can light a home and inspire a continent.

Yet the world is also moved by quieter, local heroes whose names rarely appear in headlines. In Lagos, a mother organized her neighbours to pool savings for a shared rooftop solar system. Her leadership meant that when the unreliable city grid failed, the community's children could still study under bright, clean light. In a small town along the St. Lawrence River, a Canadian teacher started an after-school energy club, guiding students to measure their school's electricity use and design their own conservation plans. Parents watched as the children became confident speakers at town meetings, teaching adults about efficiency and renewable options. In France, a mayor in a rural province defended the refurbishment of an aging nuclear plant — not out of nostalgia, but because it provided the clean, stable power that kept local farms and businesses alive. His careful listening sessions and public forums turned skepticism into a community-wide decision to keep the plant running safely, giving families both economic security and cleaner air.

Even the upheavals of migration and conflict have produced new energy champions. A Syrian refugee who settled in Germany now trains other newcomers to install rooftop solar systems, combining job creation with climate action. In the Philippines, a fisherman who once feared the rising seas now leads a cooperative that restores protective mangroves and installs small solar microgrids, proving that adaptation and mitigation can go hand in hand.

None of these people wear capes or command armies of engineers. They are scientists, diplomats, parents, teachers, refugees, and farmers who refused to wait for someone else to solve the problem. Their achievements ripple outward: A child in Toronto reads about William Kamkwamba and sketches a windmill; a family in Nairobi saves for a solar panel after hearing of the Lagos mother's success; a group of teenagers in Paris debates energy policy with newfound respect for Christiana Figueres's patience and persistence.

For families everywhere, these energy heroes transform distant concepts into living possibilities. They show that the path to a cleaner, fairer energy future is not reserved for experts or governments. It is a path that begins in kitchens, classrooms, and community halls, wherever ordinary people decide that the next generation deserves more than promises — they deserve power, light, and hope.

Why Families Need Role Models

Role models make the abstract concrete. A child hearing only that "the planet is warming" feels fear; a child hearing how William Kamkwamba built a windmill feels hope. Parents who see neighbours succeed with rooftop solar or watch a young inventor win an international prize feel less alone in their own efforts.

Stories of innovators turn despair into agency. They remind families that the energy transition is not a spectator sport. Every household can conserve, invent, or advocate. Dinner-table conversations change when children can point to real people — some their own age — proving that change is possible.

Voices That Light the Future

The story of energy is ultimately a human story. It is written in the quiet determination of researchers chasing fusion, in the songs of island poets demanding justice, and in the hands of children wiring their first solar lamp.

For families everywhere, these voices matter because they show that resilience is not just technical but profoundly human. The future will not be built only by engineers in labs or politicians in capitals. It will be built by families who, inspired by innovators, choose fairness, creativity, and courage.

Energy is not only power in wires. It is power in people — and those voices, loud and small, global and local, will light the future as surely as any reactor or turbine.

Family Action:
Celebrate the innovators who light the world — and become one. Share their stories at home, spark your own ideas, and take action that turns inspiration into impact. The next energy hero could be sitting at your table.

CHAPTER 29

Looking Ahead Together

Building a shared journey of resilience and hope

The End of the Book, the Start of the Journey

Every story eventually reaches its last chapter, but energy is not only a story we read. It is a story we live each day — in the quiet hum of a refrigerator, the glow of a streetlight, the warmth of a heater on a winter night, the solar panel that greets the dawn. As families finish this book, they are not reaching an ending but a beginning — the moment when reflection turns into action and daily choices merge into a collective force.

The journey ahead is daunting. Climate change is real. Fossil fuels still dominate. Injustices linger, and technology alone cannot erase them. Yet hope is also real. Families today hold more tools, more choices, and more collective wisdom than any generation before them. Looking ahead is not about certainty, but about courage — and courage, like energy itself, grows stronger when it is shared. Around countless kitchen tables, parents and children are already shaping tomorrow each time they talk about a new appliance, a neighbourhood solar co-op, or the dreams they hold for their grandchildren.

From Individual to Collective

Energy choices often begin at the most intimate level. A parent decides to switch to LED bulbs. A child reminds everyone to turn off lights when leaving a room. A family chooses to insulate their home or to buy an electric car. These acts may seem small, but multiplied across millions of households they can shift entire national grids and even global markets.

At the same time, no family stands alone. Neighbourhoods form energy cooperatives to buy solar panels in bulk and share battery banks. Indigenous nations co-lead transmission projects that connect distant communities to clean power and insist on consent and revenue sharing. Cities commit to ambitious climate goals that ripple outward across borders. Countries join forces to fund fusion research and offshore wind, creating networks of innovation that stretch from Arctic villages to equatorial islands.

Families are both seeds and forests: Individual actions plant the roots, and collective action becomes the canopy that sustains the whole. A single family deciding to conserve energy today is part of a vast, unseen alliance with families across oceans who are doing the same. Each small act — insulating a drafty attic, joining a local microgrid — becomes a leaf on a global tree of resilience.

Resilience as Daily Life

Resilience is often imagined as something dramatic — towering seawalls against storms or vast national grids capable of withstanding massive blackouts. Yet the truest form of resilience also lives quietly inside households and neighbourhoods, woven into daily habits that families pass down like recipes or lullabies.

In Puerto Rico, families built solar microgrids after Hurricane Maria so that schools and clinics would never again sit dark for months. Parents there talk about resilience the way they once talked about hurricane shutters: a normal part of home maintenance. In South Africa, neighbours share generators and even communal refrigerators during load-shedding, transforming hardship into solidarity and mutual care. Rural Canadian parents prepare blackout kits not out of fear but out of foresight, treating resilience as routine — no different from stocking up for winter or preparing for a new school year.

Around the world, similar stories unfold. In Bangladesh, women's cooperatives train members to maintain village solar arrays so that when cyclones strike, power and knowledge remain local. In Australia's fire-prone regions, families organize "energy-ready weekends," testing backup batteries and reviewing evacuation routes alongside neighbourhood barbecues. In urban United States apartment blocks, tenants create phone trees and pool funds for shared battery storage, ensuring that elderly residents are checked on during heatwaves.

Resilience, then, is not only about bouncing back after a disaster. It is the everyday discipline of planning, learning together, and making sure no one is left behind. Families practice it the same way they prepare for exams, weddings, or the birth of a child: with care, foresight, and love. By teaching children to keep a flashlight handy or to help a neighbour during an outage, parents show that resilience is simply another word for community.

Justice as Compass

Throughout these chapters, one truth has echoed: Energy without justice is incomplete. Clean technologies and shiny infrastructure mean little if the benefits and burdens remain unevenly shared.

Indigenous families remind the world that consent matters and that projects built on their lands must respect sovereignty. Pacific Island families, watching the ocean rise against ancestral graves, remind us that survival is a moral obligation, not a policy option. Families in coal towns warn policymakers that a transition must never abandon workers who powered past prosperity. Parents in crowded urban neighbourhoods remind us that pollution is not distributed equally; a child living beside a freeway or refinery breathes a different air than a child in a leafy suburb. Migrant families, often in the lowest-rent housing near industrial corridors, quietly pay the health costs of other people's convenience.

Justice is the compass pointing the way forward. It demands that a future be not only clean but fair, that energy systems heal old wounds instead of deepening them. A just energy plan guarantees that light reaches every home, that opportunity includes every child, and that history no longer repeats harm. Like a fair family dinner where everyone has a seat and everyone shares both the meal and the cleanup, a fair grid makes space for all — rural and urban, rich and poor, present and future generations. Without this moral orientation, even the greenest network of wires would remain unfinished.

The Role of Imagination

Imagination is not frivolous; it is fuel. Without it, no one would have envisioned solar panels glinting on rooftops, offshore wind farms stretching like white forests across the sea, or a fusion reactor attempting to bottle a star.

Children sketching windmills in classrooms, filmmakers dreaming of fusion-powered cities, musicians writing songs about rivers and sun — these acts matter. They shape how families feel about energy, and feelings shape action. A child who draws a neighbourhood powered entirely by sunlight may become the engineer or policymaker who brings that sketch to life. Parents who once read science fiction novels about limitless clean energy may carry that spark into careers or community activism decades later.

Imagination also bridges culture and technology. Storytellers in the Sahel describe solar power as "capturing the desert's heartbeat." Poets in the Arctic write of wind as an ancestor guiding the people home. These creative visions give engineers and politicians something more than blueprints: They give them purpose. When families tell stories of tomorrow around the dinner table — of homes that share power freely, of cities cooled by trees and clean grids — they are quietly drawing the first lines of that future.

To look ahead, families must imagine boldly. They must see not only what is, but what could be: grids that are both clean and fair, communities that are resilient, and children who are empowered to lead. Imagination, passed across dinner tables and bedtime stories, becomes the quiet blueprint for the future.

Hope in Global Stories

Across the world, families are already writing hopeful chapters.

Across the book, we've encountered these stories one by one. Here, they come together.

In Ontario, coal is gone and children breathe cleaner air than their grandparents imagined possible. In France, steady electricity bills reflect decades of nuclear commitment and a cultural choice for independence. In Kenya, solar lamps, rather than smoke-filled kerosene light, allow children to read long after sunset, turning evening hours into time for more easeful learning. In Germany, villages blanket rooftops with panels as symbols of local pride and self-reliance. In South Korea, students peer into fusion laboratories and talk excitedly about bottling the power of the sun. In Chile, Indigenous leaders stand at planning tables demanding fairness in every new project. In the Marshall Islands, solar panels gleam defiantly against rising seas, proof that even the most vulnerable can lead.

And there are countless quieter stories: a farmer in Vietnam who powers his irrigation pump with a single wind turbine; a grandmother in Morocco who teaches her grandchildren to maintain the family's rooftop panels; a youth collective in Brazil that turns discarded cooking oil into clean biodiesel. None of these stories are perfect. Each is a patchwork of challenges and victories. But together, they form a mosaic of progress that families everywhere can point to when doubt creeps in. Hope is not an abstraction — it is alive in homes, classrooms, and neighbourhoods across every continent.

The Family's Role in the Shared Journey

The question always returns: What can one family do? The answer is both simple and profound — participate.

Families participate when they learn, explaining energy to children in plain, patient language. They participate when they conserve, choosing efficient appliances, sealing drafts, and turning off lights. They participate when they invest in rooftop solar, join local cooperatives, or support clean energy programs. They participate when they vote for leaders who honour climate science and justice. And perhaps most importantly, they participate when they imagine, telling their children that their ideas and dreams truly matter.

No family is powerless. Each household is a voice, a choice, a spark. Together, those sparks can grow into a light bright enough to guide nations. The clean energy future will be built not only by engineers or legislators, but by countless families who decide, day after day, to move one step closer to fairness and sustainability.

Looking Ahead Together

The journey of energy is messy, complex, and unfinished. Yet it is also profoundly human, a story of connection that stretches from a single light switch to the orbit of the sun. Families are not merely consumers of electricity; they are custodians of the future, capable of shaping the path as surely as any engineer or legislator.

Looking ahead together means balancing risk and reward, honouring justice, embracing imagination, and building resilience. It means seeing nuclear and renewables not as rivals but as teammates, each offering strengths that complement the other. It means remembering the past — both the ingenuity and the mistakes — without being trapped by it.

Most of all, it means recognizing that the heart of energy is love: the love of parents for children, of communities for their land, of generations for each other. Love motivates a family to insulate a drafty house, a city to invest in public transit, a nation to fund the slow, patient science of fusion.

Families cannot choose the past, but they can shape the future. Every decision — installing a heat pump, joining a local solar co-op, teaching a child why the night sky matters — is a vote for the world to come. And when those choices are multiplied across millions of households, they become something vast: a shared journey of resilience and hope, a living testament that the clean, fair, and luminous future we dream of is already beginning to take shape.

Family Action:

Make a decision checklist with your family values, local grid facts, and the trade-offs you're willing to accept.

Appendices

Appendix A: Timeline of Nuclear Energy

This timeline is more than a list of dates. It traces the way ordinary families have lived alongside nuclear science — from kitchen-table curiosity to national policy. Each milestone was felt not only in laboratories and parliaments but also in homes, classrooms, and neighbourhoods around the world.

Early Sparks of Discovery

1896 — Invisible Rays Revealed

Henri Becquerel leaves a photographic plate in a drawer with uranium salts and notices a ghostly imprint. Around the same time, Marie and Pierre Curie spend long evenings in a cramped Paris lab, their young daughters asleep nearby, refining the concept of "radioactivity." Their work turns the atom from theory into living mystery.

1911 — The Atom Has a Nucleus

Ernest Rutherford's gold-foil experiment shows that atoms hold dense centres of positive charge. Newspaper illustrations invite families to imagine tiny solar systems inside every pebble and teacup.

Toward the Chain Reaction

1932 — Splitting the Atom in Spirit

James Chadwick discovers the neutron, a neutral particle that will later make chain reactions possible. Scientists write letters home comparing the find to "discovering a new colour."

1938 — Fission in Berlin

Otto Hahn and Fritz Strassmann observed uranium nuclei splitting apart. Lise Meitner and her nephew, Otto Frisch, exiled by Nazi persecution, explain the physics in a quiet Swedish forest, walking through snow as they realize the energy implications.

War and the First Reactors

1942 — Chicago Pile-1

Beneath the football stands of the University of Chicago, Enrico Fermi leads the first controlled, self-sustaining chain reaction. A handful of scientists celebrate with paper cups of Chianti, while the city above goes about its day unaware that a new age has begun.

1945 — From Power to Weapon

The atomic bombings of Hiroshima and Nagasaki sear the world's conscience. Japanese families experience nuclear energy first as devastation, shaping cultural memory and future skepticism.

Peaceful Power Emerges

1951 — First Nuclear Electricity

In a modest Idaho test reactor, four lightbulbs glow from atomic power. Photographs of the bulbs circle the globe, a symbol of possibility.

1954 — Obninsk, USSR

The first nuclear power station sends electricity to a working grid. Local families receive postcards proudly stamped "From the Atom."

1957 — Calder Hall, United Kingdom

The first full-scale commercial reactor opens. Schoolchildren tour the plant, and the British press hails the dawn of "Atoms for Peace."

Expansion and Debate

1960s–1970s — The Nuclear Boom

Reactors rise from France to the United States to Japan. In many towns, union wages and new schools follow. In others, protests begin, as parents worry about accidents and waste.

1973 — Oil Crisis

As fuel prices soar, countries from France to South Korea accelerate nuclear programs, seeing energy independence as family security.

Accidents and Reassessment

1979 — Three Mile Island, USA

A partial meltdown in Pennsylvania prompts evacuations and nightly television briefings. Families watch governors and engineers on their living-room TVs, learning the vocabulary of containment and coolant.

1986 — Chernobyl, USSR

An explosion and fire release radiation across Europe. Farmers in Sweden discover fallout on their sheep before Soviet authorities issue statements. The silence erodes public trust for a generation.

2011 — Fukushima Daiichi, Japan

A massive earthquake and tsunami disable cooling systems. Though radiation doses are lower than Chernobyl's, the confusion and slow communication leave Japanese families feeling betrayed and displaced.

Innovation and a Changing Climate

2010s — Second Thoughts and New Designs

Faced with climate urgency, nations revisit nuclear. Generation III+ reactors with passive safety features enter service in China and the United Arab Emirates. Ontario refurbishes aging units, ensuring another forty years of low-carbon electricity for Canadian families.

2020 — Breakthroughs in Fusion Research

Teams in Europe, the United States, and Asia report sustained high-temperature plasmas and the first laboratory instances of net energy gain. Parents show children news clips about "bottling a star," and classroom discussions buzz with excitement.

The 2020s and Beyond — A Dual Path

Small Modular Reactors (SMRs) are licensed in Canada and the United States, promising flexible, community-scale power. Private fusion startups in the United Kingdom, Japan, and Canada edge closer to commercial prototypes. Nations debate how to pair these advances with renewables to meet net-zero goals.

A Living Timeline

This story continues every day. When a new reactor switches on in Finland, when a village in Kenya debates a micro-reactor proposal, or when a child in Brazil builds a model fusion sun for a science fair, another entry belongs on this list. The timeline of nuclear energy is ultimately a timeline of human choice — of families everywhere deciding how to light their homes, protect their children, and imagine the future.

Appendix B: Global Energy Heroes

The clean energy story is written by scientists and welders, students and grandparents, diplomats and local organizers. These twelve profiles show that anyone — any family — can become part of the movement for a safer, fairer energy future.

1. Greta Thunberg — Youth as Catalyst

From a lone protest in Stockholm to a worldwide climate movement, Greta proved that a teenager with a sign can change the global conversation. Parents around the world describe their children quoting her words at the dinner table: "If she can speak up, why can't we?"

2. William Kamkwamba — The Boy Who Harnessed the Wind

As a Malawian teenager, William built a working windmill from scrap to power his family's home during a famine. His invention, later a book that was made into a film, still inspires school projects on every continent.

3. Christiana Figueres — Architect of the Paris Agreement

The Costa Rican diplomat who united nearly two hundred nations around climate action often credits her parents for teaching optimism as a moral choice. Her story shows families that diplomacy and persistence can reshape the planet's future.

4. James Hansen — The Scientist Who Spoke Up

When the NASA scientist warned the U.S. Congress about greenhouse gases in 1988, families watching on TV heard climate change explained clearly for the first time. His courage helped launch decades of climate awareness.

5. Sheila Watt-Cloutier — Voice of the Arctic

An Inuit leader from Canada, Watt-Cloutier frames climate change as a human rights issue. Her vivid stories of hunting trips and melting sea ice help families everywhere see the moral stakes of energy decisions.

6. Fatih Birol — Champion of Practical Pathways

As executive director of the International Energy Agency, Birol argues that the energy transition must protect the world's poorest. Policymakers and parents alike cite his insistence on fairness as proof that clean energy must also be just.

7. Vandana Shiva — Defender of Land and Water

The Indian environmentalist links energy, agriculture, and justice. Her campaigns for seed diversity and against destructive dams remind families that power systems must respect both people and ecosystems.

8. Bertrand Piccard — Solar Explorer

This Swiss pilot circled the globe in a solar-powered airplane, proving that clean technology can conquer even the widest oceans. Children following his journey online often wrote letters saying, "You showed us the sun can fly."

9. Leah Stokes — Scholar and Storyteller

A Canadian-American political scientist, Stokes translates complex energy policy into plain language on podcasts and in classrooms. Parents praise her work for helping teenagers understand how civic action drives real change.

10. Ayana Elizabeth Johnson — Ocean Guardian

The U.S. marine biologist and policy advocate founded the Urban Ocean Lab and co-created the *All We Can Save* anthology, showing that climate solutions must include oceans and coastal communities where millions of families live.

11. Tony Seba — Visionary of Disruption

Through research on solar, storage, and electric vehicles, this Silicon Valley thinker forecasts how fast clean technologies can outcompete fossil fuels. His optimistic models give families a sense of momentum and possibility.

12. Local Heroes Everywhere

Beyond famous names are countless unsung innovators:

a. The school principal in Kenya who organized neighbours to buy a shared solar array
b. The retired coal miner in Kentucky who now trains apprentices to install heat pumps
c. The grandmother in the Philippines who leads a village micro-hydro project so children can study at night

These quiet leaders prove that heroism is not limited to headlines. Every family can nurture its own energy hero — someone who sees a need, takes a risk, and lights the way forward.

Appendix C: Glossary of Energy Terms

Baseload Power

The steady, always-on electricity a grid needs every hour of every day. Nuclear plants and big hydro dams often provide this dependable foundation.

Battery Storage

Systems that store electricity for later use, from home batteries to giant grid installations. They soak up extra **solar** or **wind power** in the day and release it after sunset.

Carbon Capture and Storage (CCS)

Technology that traps carbon dioxide from smokestacks or even directly from the air and stores it underground so it doesn't heat the planet.

Carbon Footprint

The total greenhouse gases released by a person, family, or activity. Turning off lights or driving less shrinks this "footprint."

Carbon Neutral / Net-Zero

Balancing any carbon emissions with an equal amount removed from the atmosphere so that overall warming does not increase.

Climate Resilience

The ability of communities and systems to withstand and recover from climate impacts — like a house built to ride out a storm.

Community Solar / Wind

Projects where neighbours jointly own or subscribe to a renewable power installation and share the electricity or revenue.

Decommissioning

The careful process of retiring a nuclear plant, wind farm, or other energy facility at the end of its life. Includes dismantling equipment, managing waste, and restoring the site so it can be safely reused.

Demand Response

Programs that reward families or businesses for shifting energy use to off-peak hours, helping the grid stay balanced.

Distributed Energy

Power produced close to where it's used — rooftop solar, backyard wind, local batteries — rather than at distant plants.

Dunkelflaute (German, "dark doldrums")

Long winter stretches with little sun or wind, a challenge for renewable-heavy grids.

Electric Vehicle (EV)

A car, bus, or bike powered by electricity instead of gasoline or diesel.

Energy Efficiency

Using less energy to do the same job — LED bulbs that shine as brightly as old ones but sip electricity.

Energy Justice

Fair distribution of energy's benefits and burdens: who enjoys clean power, who breathes the pollution, and who has a voice in decisions.

Energy Literacy

Understanding where power comes from, how it's used, and how choices affect both the planet and the family budget.

Fission

Splitting heavy atoms (like uranium) to release energy — how today's nuclear plants make electricity.

Fusion

Joining light atoms (like hydrogen) to release enormous energy — the reaction that powers the **sun** and the dream of future reactors.

Geothermal Energy

Heat from beneath the earth's surface used to generate electricity or warm buildings.

Gigawatt / Megawatt / Kilowatt

Units of electrical power. One gigawatt can power roughly a million homes; a kilowatt is about what a space heater uses.

Grid

The network of wires, substations, and control centres that moves electricity from where it's made to where families use it.

Hazardous Waste

By-products from energy production — such as spent nuclear fuel or coal ash — that require careful, long-term handling to protect people and ecosystems.

Hydrogen Fuel

A clean energy carrier that, when burned or used in a fuel cell, produces only water. It can store energy or power vehicles.

Intermittency

The natural ups and downs of solar and wind output — cloudy days, calm nights — that make storage and backup essential.

Microgrid

A small, local energy network that can run independently during a blackout — like a neighbourhood with its own mini–power plant.

Net Metering

A billing system that credits households for excess electricity their solar panels send back to the grid.

Peak Load

The highest level of electricity demand in a given period, such as early evening when lights, ovens, and TVs all switch on.

Photovoltaic (PV) Cell

The basic unit of a solar panel, turning sunlight directly into electricity.

Power Purchase Agreement (PPA)

A long-term contract in which a customer (often a city or school) agrees to buy power from a renewable project at a set price.

Radiation

Energy travelling as waves or particles. In this book, it often means the controlled release from radioactive materials — monitored carefully in medicine and nuclear power.

Remediation

Cleaning up and restoring land or water that has been contaminated — whether from old mines, oil spills, or decommissioned nuclear sites — so families can use it safely again.

Renewables

Sources of energy that don't run out on human timescales: sunlight, wind, moving water, geothermal heat.

Smart Grid

An upgraded electricity network that uses sensors and software to match supply with demand, detect outages, and integrate renewables.

Small Modular Reactor (SMR)

A new generation of compact nuclear reactors that can be built in factories and assembled on-site, offering flexible, lower-cost nuclear power.

Social Cost of Carbon

An estimate of the economic damages caused by emitting one tonne of carbon dioxide — used by policymakers to weigh the real cost of fossil fuels.

Time-of-Use Pricing

Electricity rates that change by the hour, encouraging families to run dishwashers or charge cars when clean power is plentiful and cheap.

Transmission Lines

High-voltage cables that carry electricity long distances from power plants to local distribution networks.

Waste (Nuclear or Industrial)

Any material left after energy production that must be stored or treated safely, from spent nuclear fuel to chemical by-products of fossil fuels. Modern nuclear waste is carefully contained and monitored for decades or centuries.

Waste Heat / Cogeneration

Capturing heat that would otherwise be lost — such as from factory furnaces — to generate extra electricity or to warm nearby buildings.

Appendix D: Recommended Resources

These resources can help families keep learning, exploring, and participating in the clean energy future you've been reading about. They were selected to reflect the themes of the book — *family conversations, energy literacy, justice, imagination, and action* — and all are available in English, with many offering translations or subtitles.

Books and Long-Form Reading

1. ***The Grid*** by Gretchen Bakke — A vivid, approachable history of the electric grid and why resilience matters for every household.
2. ***Sustainable Energy — Without the Hot Air*** by David MacKay — Clear explanations of how different energy sources add up, written for curious lay readers.
3. ***Power to the People*** by Astrid Kander et al. — A global look at how energy and society shape each other across centuries.
4. ***The Boy Who Harnessed the Wind*** by William Kamkwamba — The true story of a Malawian teenager who built a windmill to save his village, perfect for family reading or teen book clubs.

For Parents and Educators

1. **U.S. Energy Information Administration, "Energy Kids"** — Interactive games and lesson plans explaining everything from solar panels to nuclear reactors.
2. **National Renewable Energy Laboratory (NREL) Learning Resources** — Free activities and science kits for classrooms and community groups.

3. **Canadian Nuclear Safety Commission Educational Portal** — Videos, infographics, and FAQs to help families understand how reactors are built and monitored.
4. **UNESCO Education for Sustainable Development** — Global curriculum guides and case studies for teachers who want to integrate climate and energy into everyday subjects.

For Teens and Young Adults

1. **International Atomic Energy Agency, "Nuclear Explained"** — Short, plain-language explainers about fission, fusion, and radiation safety.
2. **MIT Climate Portal** — Podcasts, Q&A sessions, and easy-to-share explainers on climate science and energy technology.
3. **SolarSPELL Digital Library** — Open-access solar-powered learning kits used in schools from the Pacific Islands to Africa.

Films and Documentaries

1. ***WALL-E***, ***Moana***, and ***Princess Mononoke*** — Family-friendly films that spark conversation about stewardship, balance, and energy use.
2. ***Chernobyl*** (HBO miniseries) — A gripping dramatization of the 1986 accident, useful for older teens and adults discussing trust and transparency.
3. ***Before the Flood*** and ***2040*** — Inspiring documentaries about renewable solutions and hopeful futures.
4. ***The Boy Who Harnessed the Wind*** (Netflix) — Movie adaptation of William Kamkwamba's memoir, celebrating youth innovation.

Community and Participation

1. **Mothers Out Front** — Parent-led movement advocating for clean energy and climate action at the local level.
2. **Citizens' Climate Lobby** — Family-friendly advocacy network supporting fair climate and energy policies.
3. **Indigenous Clean Energy Network** (Canada) — Programs and training opportunities for Indigenous families and youth.
4. **Transition Network** — Global grassroots communities building local resilience, from community gardens to microgrids.

Reliable News and Data

1. **International Energy Agency (IEA)** — Up-to-date global statistics and policy reports.
2. **Our World in Data: Energy** — Beautifully visualized graphs and country comparisons, easy for students to explore.
3. **World Nuclear Association** — Balanced facts on nuclear energy, safety, and new reactor technologies.
4. **REN21, *Renewables Global Status Report*** — Annual report tracking the growth of solar, wind, and other clean power sources.

Hands-On and At-Home Activities

1. Build a **pizza-box solar oven** (instructions from many science museums) and host a solar-baked cookie night.
2. Start a **family energy diary**: Track electricity use, then brainstorm ways to cut it by 10 percent over a month.
3. Visit a **local power plant**, renewable farm, or science centre — many utility companies offer free tours for families.

These resources are invitations, not assignments. Choose the ones that match your family's interests: a bedtime story, a weekend experiment, a policy briefing for older teens, or a community meeting. Each can help keep the conversation alive long after you turn the final page of this book — ensuring that your family remains not just a consumer of energy, but a creator of a cleaner, fairer future.

Appendix E: Family Discussion, Action Guide, and Questions to Ask

This appendix combines a Family Discussion Guide and a Family Action Checklist to help families reflect on energy choices and take meaningful steps toward a cleaner, more informed future.

Family Discussion Guide

Bringing the Energy Conversation Home

Use these questions to spark thoughtful conversations around the dinner table, during a walk, or at a family meeting. You don't need to answer them all at once — pick a few that resonate and revisit others over time.

1. **Energy Awareness**

Where does our electricity come from?

Have we ever experienced a power outage? How did it affect us?

What appliances or habits use the most energy in our home?

2. **Values and Priorities**

What matters most to us when it comes to energy: cost, safety, climate, reliability, or something else?

How do we feel about the trade-offs between different energy sources?

Are we more concerned about risks today or long-term impacts on future generations?

3. **Climate and Community**

How is climate change affecting our region or people we know?

What role should our community play in reducing emissions?

How do we feel about nuclear energy as part of the solution?

4. **Trust and Decision-Making**

Who do we trust to make energy decisions — scientists, governments, communities?

What would help us feel more confident in those decisions?

How can we make our voices heard?

5. **Looking Ahead**

What kind of energy future do we want for our children or grandchildren?

What small steps could we take to align our home with that vision?

What would we like to learn more about?

Family Action Checklist

From Awareness to Action

Choose a few actions that feel right for your family. Every step counts.

1. Learn Together

Watch a documentary or video about energy or climate change.

Visit a local power plant, science centre, or energy museum.

Read a book or article together about nuclear or renewable energy.

Invite a guest speaker (e.g., engineer, teacher, elder) to a school or community group.

2. Make Your Home Smarter

Switch to LED bulbs or smart thermostats.

Apply for energy efficiency rebates or home audits.

Track your electricity use for a week and discuss what you find.

Explore options for green energy plans or rooftop solar.

3. Get Involved

Attend a town hall or public consultation on energy or climate.

Write to your local representative with a question or concern.
Join a local environmental or energy justice group.
Start a school or neighbourhood energy project.

4. Support Future Careers
Encourage kids to explore STEM or energy-related careers.
Look into scholarships, internships, or mentorships in clean energy.
Share stories of diverse role models in nuclear and renewable fields.

5. Think Big, Act Local
Talk to neighbours about energy choices and climate resilience.
Support policies that promote clean, reliable, and fair energy.
Reflect on how your family's choices connect to global challenges — and solutions.

Questions to Ask Your Utility or Local Government

Where does our electricity come from, and how has that changed over the past ten years?

This helps uncover whether your region is shifting toward cleaner or more reliable sources — or still relying heavily on fossil fuels.

Is nuclear energy part of our current or future energy plans?

Some communities are exploring nuclear for the first time, while others may be phasing it out. This question opens a broader conversation about reliability, emissions, and safety.

What programs exist to help families save energy or reduce bills — especially for renters or low-income households?

You may discover rebates, energy audits, or support programs that aren't well advertised.

How is our community preparing for more extreme weather, and is the grid ready to handle it?

A good way to surface conversations about grid upgrades, battery storage, emergency planning, and local energy resilience.

What steps are being taken to reduce emissions from electricity generation?

This invites dialogue about timelines, targets, and trade-offs.

Are Indigenous communities, local residents, and other stakeholders being consulted about new energy projects?

Energy justice includes who is heard, not just what is built. This question shows that fairness matters.

Can we choose where our electricity comes from — or opt in to clean power programs?

Some utility companies let customers pay slightly more for renewable energy or join community solar projects. Others may offer rooftop solar programs or time-of-use billing to help reduce strain on the system.

Are there any new technologies — like Small Modular Reactors or energy storage — being considered for our area?

This can spark conversations about innovation, timelines, public consultation, and potential partnerships.

What opportunities exist for public input on local energy or infrastructure decisions?

Sometimes utility companies hold community forums or publish draft energy plans for review. Asking can help you get involved early.

How are schools, hospitals, and other public buildings being made more energy-efficient?

A smart way to push for transparency — and for the institutions we rely on to lead by example.

Is our community working on a climate or energy transition plan? If so, how can residents contribute?

Many towns and cities are developing long-term strategies. This question shows you're engaged and want to help.

What happens during a blackout or supply emergency? Who's most vulnerable — and how are they being protected?

A caring question that also raises equity concerns, especially in extreme heat or cold.

Bibliography

This book draws on a blend of scientific research, international energy reports, historical records, and contemporary journalism. Below is a section-by-section guide to key sources that families, students, and policymakers can consult for deeper reading.

Introduction

1. International Energy Agency (IEA). *World Energy Outlook* — for global statistics on electricity access, costs, and projections.
2. Intergovernmental Panel on Climate Change (IPCC). *Sixth Assessment Report* — foundational climate science and emissions data.
3. U.S. Department of Energy and Natural Resources Canada — national energy mix and outage data.

Part I: Families, Energy, and Everyday Life

1. World Health Organization. "Radiation: Health Effects."
2. Canadian Nuclear Safety Commission. *Radiation and Your Health.*
3. U.S. Energy Information Administration (EIA). *How Electricity Gets to You.*
4. Case studies on family energy resilience from Puerto Rico and South Africa drawn from reporting in *The Guardian* and *National Geographic.*

Part II: Energy Choices and Trade-Offs

1. IEA. *Renewables 2023 Market Report.*
2. U.S. National Renewable Energy Laboratory (NREL) — cost comparisons of wind, solar, and storage.
3. International Monetary Fund and World Bank reports on global energy subsidies and fossil-fuel economics.

4. European Union Agency for the Cooperation of Energy Regulators — data on the 2022—2023 gas price crisis.

Part III: The Role of Nuclear in a Changing World

1. World Nuclear Association (WNA). Country profiles for France, Canada, South Korea, and the UAE.

2. United Nations Declaration on the Rights of Indigenous Peoples (UNDRIP) — principles guiding consent and participation.

3. Historical documentation on Chernobyl (1986) and Fukushima (2011) from the International Atomic Energy Agency (IAEA) and peer-reviewed health studies.

4. Academic works such as Gabrielle Hecht's *The Radiance of France* and Kate Brown's *Manual for Survival* for cultural perspectives.

Part IV: Looking Ahead

1. International Labour Organization (ILO). *Global Employment Trends for the Energy Sector.*

2. ITER Organization and U.S. National Ignition Facility — fusion research updates.

3. Small Modular Reactor design data from the IAEA and the U.S. Nuclear Regulatory Commission.

4. Reports on just transitions from the International Institute for Sustainable Development and the World Resources Institute.

Part V: Energy, Culture, and Imagination

1. Analyses of music, film, and literature drawn from journals such as *Environmental Humanities* and *Energy Research & Social Science.*

2. Primary cultural sources: traditional coal-mining ballads, protest songs archived by the American Folklife Center, and contemporary climate-movement music.

3. Film references including *Godzilla* (1954), *WALL-E* (2008), and *Moana* (2016).

4. Interviews and field reports from global artists and Indigenous storytellers, published in outlets like *Orion* and *Emergence Magazine*.

Appendices

1. Historical nuclear milestones verified through the IAEA Nuclear Data Section and U.S. Department of Energy archives.

2. Interview excerpts compiled from public forums, parent-teen dialogues shared during community energy workshops, and publicly available policy interviews.

3. Biographical information for global energy heroes from national academies of science, Nobel Prize archives, and reputable news profiles.

These references are intended to guide readers toward the same reliable, publicly available materials that informed the manuscript's research and toward real-world articles. They are organized to match the table of contents so families, students, and educators can easily continue their own exploration of the energy stories shaping our shared future.

Figures Citation

Fig. 1: Author's portrayal of boardroom and kitchen-table discussions of energy and energy bills
Photos by Vlada Karpovich and Mikhail Nilov from Pexels

Fig. 2: Canada's electricity mix vs the global electricity mix, 2023
Statistics Canada, Electricity generation, by type of electricity producer (2023)
Ember, *Global Electricity Review 2024*

Fig. 3: Average number of smog advisory days in Ontario before, during, and after the coal phase-out
Bars show the average number of Ontario-wide smog advisory days per year for three periods: pre–coal phase-out (2003–2007), transition (2008–2013), and post–coal phase-out (2014). Data are from Ontario Ministry of the Environment, Conservation and Parks (MECP), *Air Quality in Ontario* annual reports, 2003–2014. Coal-fired electricity generation was fully eliminated in Ontario in 2014, coinciding with the elimination of smog advisories that year.

Fig. 4: Comparison of typical radiation doses from common sources
Approximate effective doses (μSv) are shown for representative everyday exposures. Banana dose reflects ingestion of potassium-40; granite countertop represents typical annual indoor terrestrial background; flight dose corresponds to a single commercial transatlantic flight (New York—London); nuclear plant value reflects typical measured annual public exposure near nuclear facilities, not regulatory limits.

Fig. 5: Average sources of radiation exposure in the global population
The figure shows the approximate contribution of major radiation sources to the average annual effective dose. Medical exposure and indoor radon dominate population dose, while contributions from nuclear power generation are negligible (less than 0.1 percent). Values

are based on assessments by UNSCEAR and related national radiation protection bodies.

Fig. 6: Simplified pressurized water reactor (PWR), nuclear power plant, and electricity delivery pathway
Thermal energy produced in the reactor pressure vessel is transferred via a closed primary coolant loop to a steam generator, where steam is produced in a separate secondary loop. Steam drives a turbine connected to an electric generator. Electrical power is stepped up for high-voltage transmission, then stepped down for local distribution to end users. Diagram is schematic and not to scale. Diagram is based on standard PWR designs described by the International Atomic Energy Agency (IAEA) and U.S. Nuclear Regulatory Commission (NRC).

Fig. 7: The Grid: How the electricity grid works
Electricity is generated at power plants using a variety of energy sources, including water, wind, sunlight, and heat. It is carried over long distances by high-voltage power lines, adjusted at substations, and delivered through local power lines to homes and businesses. The diagram is simplified and not to scale. This is an author-generated schematic based on standard electric power system descriptions (U.S. Department of Energy, International Energy Agency).

Fig. 8: From consumer to participant
In the past, electricity felt distant: Homes used power, paid a bill, and rarely thought about where energy came from. Today, homes can generate electricity with solar panels, store it in batteries, use it in vehicles, share it with neighbours, and stay connected to the wider grid — becoming active participants in the energy system.

Fig. 9: Residential electricity prices in selected regions
Monthly average residential electricity prices are shown for the United States (seasonal variation) alongside regulated residential tariffs for France and Quebec (flatlines). While price levels shift between years, the under-

lying pattern remains consistent: Regulated systems exhibit greater price stability than market-exposed systems.

United States monthly residential prices come from the U.S. Energy Information Administration (EIA), *Electric Power Monthly*. French regulated residential tariff benchmarks are based on Eurostat's national tariff publications. Data for Quebec is based on Hydro-Québec's Rate D residential tariffs.

Fig. 10: Illustrative share of households experiencing energy poverty in selected countries

Percentages shown are indicative estimates drawn from national surveys and international assessments using country-specific definitions of energy poverty (such as inability to afford adequate energy services). The figure is illustrative and intended to highlight broad global disparities rather than provide a harmonized or exhaustive comparison. Data is culled from the International Energy Agency (IEA), World Bank, Eurostat, national statistical agencies, and household energy surveys.

Fig. 11: Canada's illustrative monthly residential electricity costs by province (1,000 kWh/month, 2023)

Hydro-Quebec and provincial utility tariffs.

Fig. 12: Global electricity mix, 2023

StatsCAN.

Fig. 13: Fossil fuels: Familiar benefits, hidden costs

Graphic is author-generated based on information from various sources.

Fig. 14: Nuclear power capacity by country (top 10, 2024): net installed capacity of reactors in operation

IAEA PRIS — (Operational reactors by country, net electrical capacity)

Fig. 15: Indigenous-led clean energy projects around the world

The map highlights representative Indigenous-led renewable energy initiatives, including solar, wind, hydro, and geothermal projects, selected to illustrate a range of geographic, technological, and community contexts. The figure is illustrative and not a comprehensive inventory of projects. To create the map, the author synthesized publicly documented Indigenous-led renewable energy initiatives. The author consulted the International Energy Agency (IEA), the International Renewable Energy Agency (IRENA), the Indigenous Clean Energy Network, and national and regional Indigenous energy organizations.

Fig. 16: Public support for nuclear power for selected countries
Bars show the share of survey respondents who report supporting the use of nuclear power to generate electricity. Results are drawn from a multinational public opinion survey conducted between November 2022 and January 2023. Percentages reflect national samples and are shown for comparison across countries. Data comes from the Potential Energy Coalition's *The World Wants New Nuclear* multinational survey fielded by Dynata, November 2022—January 2023.

Fig. 17: Fission vs fusion simplified
Illustration created by author for explanatory purposes.

Fig. 18: Selected Small Modular Reactor (SMR) projects worldwide
Map created by the author illustrating the geographic distribution of representative operating, licensed, and planned SMR projects in North America, Europe, and Asia.

Fig. 19: Scale comparison: Conventional reactors vs SMRs
Author-generated image.

Fig. 20: Nuclear energy on screen: Fear and fascination over time.
Selected films and series illustrating how popular media has portrayed nuclear themes from the early Cold War to the present, reflecting shifting cultural anxieties—from atomic fear and accident narratives to historical reckoning and ethical responsibility

About the Author

Dr. Peter Ozemoyah's relationship with energy began long before his career in engineering. Growing up in Nigeria, he experienced first-hand what it meant for families to plan their days around the rhythm of electricity — moments of bright possibility, and moments of sudden darkness. Those early experiences, etched into childhood evenings lit by candles and conversation, sparked a lifelong desire to understand how energy shapes the safety, dignity, and hope of everyday people.

That path carried him across continents — from studying chemical engineering in Nigeria and the United States to teaching, leading, and delivering nuclear and energy projects in Canada. Over more than thirty-five years, he became known not only as a senior executive and engineer, but as someone who could translate technical worlds into human ones. At Tyne Engineering, where he eventually became Vice President, he worked on technologies designed to protect communities and strengthen resilience. Later, he was invited to lead the Outreach and Stakeholder Relations at the University Network of Excellence in Nuclear Engineering (UNENE), helping guide the next generation of nuclear professionals.

But titles never fully captured his motivation.

For Dr. Ozemoyah, energy has always been a family story — about warmth in winter, safety during storms, fairness of access, and the quiet comfort of knowing loved ones are protected. His service on boards across Canada and internationally, along with recognition as one of Canada's most influential Black leaders, reflects a career devoted not only to innovation, but to inclusion, education, and community care.

This book grows out of that lifelong journey: from a boy watching the lights flicker, to a father answering his children's questions during a winter blackout, to a global energy professional determined to help families everywhere understand the forces shaping their daily lives.

He believes every household deserves clarity, confidence, and a seat at the table where energy decisions are made.

Dr. Ozemoyah lives in the Greater Toronto Area, where energy conversations often begin — just as they do in this book — around his own kitchen table.

www.ingramcontent.com/pod-product-compliance
Lightning Source LLC
LaVergne TN
LVHW022159200926
848481LV00006B/1785